T0214001

EAI/Springer Innovations in Communication and Computing

Series editor

Imrich Chlamtac, European Alliance for Innovation, Ghent, Belgium

The impact of information technologies is creating a new world yet not fully understood. The extent and speed of economic, life style and social changes already perceived in everyday life is hard to estimate without understanding the technological driving forces behind it. This series presents contributed volumes featuring the latest research and development in the various information engineering technologies that play a key role in this process.

The range of topics, focusing primarily on communications and computing engineering include, but are not limited to, wireless networks; mobile communication; design and learning; gaming; interaction; e-health and pervasive healthcare; energy management; smart grids; internet of things; cognitive radio networks; computation; cloud computing; ubiquitous connectivity, and in mode general smart living, smart cities, Internet of Things and more. The series publishes a combination of expanded papers selected from hosted and sponsored European Alliance for Innovation (EAI) conferences that present cutting edge, global research as well as provide new perspectives on traditional related engineering fields. This content, complemented with open calls for contribution of book titles and individual chapters, together maintain Springer's and EAI's high standards of academic excellence. The audience for the books consists of researchers, industry professionals, advanced level students as well as practitioners in related fields of activity include information and communication specialists, security experts, economists, urban planners, doctors, and in general representatives in all those walks of life affected ad contributing to the information revolution.

Indexing: This series is indexed in Scopus, Ei Compendex, and zbMATH.

About EAI

EAI is a grassroots member organization initiated through cooperation between businesses, public, private and government organizations to address the global challenges of Europe's future competitiveness and link the European Research community with its counterparts around the globe. EAI reaches out to hundreds of thousands of individual subscribers on all continents and collaborates with an institutional member base including Fortune 500 companies, government organizations, and educational institutions, provide a free research and innovation platform.

Through its open free membership model EAI promotes a new research and innovation culture based on collaboration, connectivity and recognition of excellence by community.

More information about this series at http://www.springer.com/series/15427

Pethuru Raj • Ashutosh Kumar Dubey
Abhishek Kumar • Pramod Singh Rathore
Editors

Blockchain, Artificial Intelligence, and the Internet of Things

Possibilities and Opportunities

 Springer

Editors
Pethuru Raj
Site Reliability Engineering Division
Avana Building
Reliance Jio Information
communication Ltd.
Bangalore, Karnataka, India

Abhishek Kumar
Chitkara University
Rajpura, Punjab, India

Ashutosh Kumar Dubey
Institute of Engineering and Technology
Chitkara University
Rajpura, Punjab, India

Pramod Singh Rathore
Maharshi Dayanand Saraswati University
Ajmer, India

ISSN 2522-8595　　　　　　　　ISSN 2522-8609　(electronic)
EAI/Springer Innovations in Communication and Computing
ISBN 978-3-030-77639-8　　　　　ISBN 978-3-030-77637-4　(eBook)
https://doi.org/10.1007/978-3-030-77637-4

This Springer imprint is published by the registered company Springer Nature Switzerland AG
The registered company address is: Gewerbestrasse 11, 6330 Cham, Switzerland

Preface

The cool convergence of artificial intelligence (AI), the Internet of Things (IoT), and blockchain technologies is being seen as a paradigm shift for global businesses, IT organizations, service providers, governments, and common people also in artistically enabling real digital transformation. These path-breaking technologies have laid down a stimulating foundation for visualizing and realizing a myriad of breakthrough business workloads and IT services. Individually and collectively, these have gained an inherent power to create a greater impact on the human society. In short, the technological evolutions and revolutions are making deeper and decisive implications towards knowledge society.

The IoT paradigm has led to the grand realization of scores of digitized entities and connected devices, which, in turn, come handy in producing and sustaining connected environments such as connected, homes, hotels, and hospitals. AI has the wherewithal to meticulously process all sorts of IoT data to extract actionable insights in time. The knowledge discovery and dissemination get automated and accelerated through the smart leverage of AI algorithms, frameworks, platforms, accelerators, and libraries. Precisely speaking, AI has garnered the much-needed strength to augment the brewing idea of data-driven insights and insights-driven decisions and actions for all. AI is to replicate the widely celebrated human intelligence into everyday devices (industry machineries, defense equipment, medical instruments, kitchen utilities and ware, information appliances, portables, implantable, handhelds, wearables, and nomadic and fixed systems). Such an AI-inspired empowerment leads to the realization of intelligent systems, applications, networks, services, and environments in and around us. The IoT-induced connected entities into smarter systems.

Finally, the recent emergence of blockchain technology has brought in a number of innovations and disruptions. The aspect of digital transformation is being simplified and speeded up with the unique contributions of blockchain. The blockchain paradigm has established a sparkling platform for envisaging newer possibilities and opportunities. As discussed above, the IoT phenomenon has penetrated into every domain these days. However, there are a few critical challenges and concerns being associated with the flourishing IoT idea. Especially the security of IoT devices and data and the scalability of IoT systems are being presented as the most abhorrent limitations. Blockchain is being seen as a silver bullet in surmounting these critical issues once for all. On the other hand, there are a few noteworthy limitations of blockchain systems. These widely reported issues are being ingeniously solved through AI-induced automated data analytics. Thus, each technology is supplementing each other in bringing forth highly sophisticated applications and state-of-the-art services not only for enterprises but also for commoners.

This book is primarily planned and prepared with the sole aim of conveying and clarifying the nitty-gritty of these futuristic and flexible technologies in building, deploying, and delivering digital life applications. Researchers in academic institutions and IT companies are striving hard in pinpointing the drawbacks of each of these technologies and they articulate and accentuate how they can be overcome through the fusion of these technologies. The chapters are well developed and incorporated, and provide a lot of detail so that our esteemed readers get a good grasp of the technologies and their strategic contributions.

This book is organized into ten chapters.

Chapter 1 discusses the latest trends in blockchain technology and its applications, and focuses on certain features of this disruptive technology which can transform traditional business practices in the near future. There are many researchers working on this field for more than 10 years and published their articles in highly reputed journals. Also, many reports and white papers are available. Therefore, a systematic collection of those articles and presenting a review of those literatures is highly useful for upcoming researchers. In this chapter, a detailed classification of blockchain-based applications from different domains such as land registration, healthcare, IoT, security, and latest areas for research is presented.

Chapter 2 tries to show that the traffic analysis module uses already provided surveillance cameras to analyze the parameters by using artificial intelligence and optical character recognition through CCTV cameras. Each lane on the road is distinguished into three or more segments according to the scenario. These lines are used to determine the traffic density. Whenever vehicles hide a part of the line in a lane, we can estimate the Traffic density concerning time, where the camera frames correspond to time. In this manner, we can generate real-time traffic data and implement a time sequence for the signal to clear off the traffic. The scope can be further widened by using this system together with INTELLIGENT TRANSPORT SYSTEMS. The module is set so that it will be trained to get data from processed datasets. The advanced structure provides an embedding of a GPS module in a vehicle which will be able to identify emergency vehicles and can also minimize vehicle theft. The module with the GPS sensor makes a good accurate score on the prediction. For the implementation of this system, video feed from the CCTV cameras is used to detect the amount of traffic density in each lane using OpenCV.

Chapter 3 explains metrics to assess the performance in the use of project management methodologies, and the first results on a small number of use cases. From bibliographic sources, a set of new metrics are defined to include those evaluations that up to now are subjective or non-existent. As part of the scope are a detailed presentation of some of the metrics and their application to a case study, along with some statistics. The applicability of the proposal to real cases is also analyzed. The metrics details, their complete listing and the functional prototype are not discussed in this chapter.

Chapter 4 – This chapter has explained about the Bitcoin digital money. Blockchain is expected to be expanded to a broad spectrum of financial applications as well as to other industries, such as delivery and shared economies. This larger deployment calls for a range of technical challenges to be tackled through the clouding of data privacy protections and improved processing speed, and the Fujitsu Laboratory is working on several related R&D projects. This research outlines blockchain technologies and sample applications, describes encryption technology in the business sense, and examines Fujitsu's communications activities.

Chapter 5 deals with the concept that it is difficult to guarantee zero downtime even if the backhaul network is optimally deployed. An automated system that could predict the downtime and take suitable steps to ensure high availability (HA) of resources could protect businesses from losing data at a critical time. Such an automated model is put forward as it would guarantee zero downtime deployment without comprising the security of end devices. The emphasis is on designing an effective solution by improvising the current strategies for this deployment challenge. The proposed architecture is evaluated through numerous tests and by visualization of the results obtained; one can mitigate the planned and unplanned downtime to ensure continuous operational efficiency of services for businesses.

Chapter 6 explains strategies that can be used to efficiently include cloud storage with protection, confidentiality, and privacy. Privacy and power aspects play a significant role in the transferring of information through IoT due to restricted interconnected systems power and processing, i.e., processing and storage resources. Whether fraudulent or unintentional, there are possible real-world effects of data intrusion in an IoT system. This chapter has discussed the potential to combine blockchain technology and software defined networking (SDN) to alleviate some of the problems. Particularly, using cluster architecture with a modern routing protocol, we suggest a stable and energy-saving blockchain-enabled SDN architecture for IoT. The experimental results suggest that the cluster structure–based routing and packet-oriented routing have lower latency and power consumption and are more robust than the above routing protocols. The cluster structure–based routing protocol has improved performance than other routing protocols with respect to performance, power savings, and robustness.

Chapter 7 discusses the concept of client server model for healthcare. The personal health record and electronic health record play a vital part in helping patients and healthcare providers recover data more efficiently. However, it is difficult to achieve a cohesive understanding of health information that is spread through multiple healthcare professionals. Health reports are typically distributed in various locations in particular and are not synchronized. Electronic health records (EHRs) have developed a common way to preserve and track facts in healthcare for patients. EHR is securely housed through using client-server model whereby each patient maintains patient data stewardship. A patient's records are distributed using focused individual servers between various hospitals.

Chapter 8 discusses primarily the goal of VANETs, which is to assist communication between vehicles and roadway units (RSUs) through vehicle-to-RSU (V2R) and vehicle-to-vehicle (V2V) networks. Due to the growing number of smart cities across the globe and innovation in the field of technology, VANET apps have a tremendous potential for growth. Road accidents and traffic congestion are growing daily because of the rapid increase in the day-to-day use of vehicles. In order to prevent emergencies and resolve the problem of vehicle congestion, coordination between vehicles is important.

Chapter 9 concerns the use of blockchains in overseeing and sharing electronic health and medical records to permit patients, emergency clinics, centers, and other stakeholders to share information among themselves, and increase interoperability. In spite of the fact that utilization of blockchains may decrease redundancy and give guardians reliable records about their patients, it accompanies few difficulties which could encroach patients' security, or possibly bargain the entire system of partners.

Chapter 10 – This chapter has focused on the various benefits that an IoT organization typically offers some benefits, still, Current Centralized Architecture (CCA) introduces the Involving various problems like privacy, safety, data integrity, transparency and single point of failure. An obstacle in the way of IoT applications and challenges in the future developments. It is one of the moving IoT toward distributed ledger technologies is one of the moving into IoT to the correct choice to resolve these problems. The popular type and common of distributed ledger technologies is the blockchain. Integrating the IoT with blockchain innovation can bring endless advantages. They depend on centralized models that present another assortment of specialized impediments to oversee them all around the world. Engineering for refereeing jobs and authorizations in IoT.

Bangalore, Karnataka, India   Pethuru Raj
Rajpura, Punjab, India   Ashutosh Kumar Dubey
Rajpura, Punjab, India   Abhishek Kumar
Ajmer, India   Pramod Singh Rathore

Contents

A Comprehensive Survey on Blockchain and Cryptocurrency Technologies: Approaches, Challenges, and Opportunities

K. R. Jothi ⓘ **and S. Oswalt Manoj** ⓘ

1 Introduction

The blockchain is a disruptive, state-of-the-art technology that operates on the decentralization principle and consists of a network of computing peer-to-peer nodes that verifies and validates every digital transaction in a manner that allows only one true copy of the ledger to exist. A blockchain is fundamentally a disseminated database of records or an open record, everything being equivalent or automated events that have been executed and shared among sharing gatherings. Each transaction in the open record is checked by the understanding of a larger piece of the individuals in the structure. Once entered, information can never be erased. The blockchain contains a certain and obvious record of every single trade anytime made. The basic blockchain architecture is given in Fig. 1. Bitcoin, the decentralized e-cash, is the most standard model that uses blockchain development. The automated money Bitcoin itself is significantly questionable, yet the essential blockchain advancement has worked impeccably and found the vast extent of employment in both budgetary and noncash-related domains.

The applications of blockchain can be implemented in almost any field out there, be it artificial intelligence or entertainment. Blockchain appears to be a one-stop destination to numerous business problems, be its steely security protocols, proficiency, and mature networking with 100% accountability, or be it the disintermediation and problem-solving in an inventive way to bring about a change in the way people utilize their technology. Yes, the distributive ledger does it all.

K. R. Jothi (✉)
School of Computer Science and Engineering, Vellore Institute of Technology, Vellore, India

S. Oswalt Manoj
Department of Computer Science and Business System, Sri Krishna College of Engineering and Technology, Coimbatore, India

© Springer Nature Switzerland AG 2022
P. Raj et al. (eds.), *Blockchain, Artificial Intelligence, and the Internet of Things*,
EAI/Springer Innovations in Communication and Computing,
https://doi.org/10.1007/978-3-030-77637-4_1

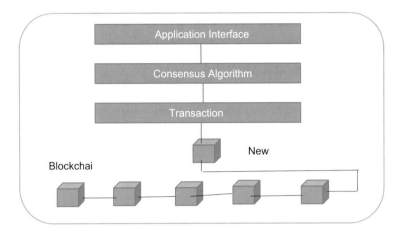

Fig. 1 Blockchain architecture

Fig. 2 Types of blockchain Networksnetworks

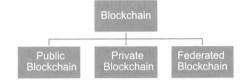

Blockchain advancement is discovering applications in a vast extent of locales, both cash-related and nonbudgetary. Cash-related associations and banks no longer watch blockchain development as a threat to customary plans of action.

Nonfinancial application openings are also ceaseless. We can envision putting check of essence of each and every legitimate report, prosperity records, unwavering quality portions in the music business, lawful official and private assurances, and marriage licenses in the blockchain. There are three different types of blockchain networks, as given in Fig. 2. They are (1) public blockchain, (2) private blockchain, and (3) federated blockchain. As the name indicates, the public blockchain is available for the public to view the transactions, whereas in the private blockchain, the transactions are accessed only by the internal members within the organization. In the federated blockchain, a few transactions are publicly visible and a few transactions are private.

By taking care of the one-of-a-kind sign of the modernized asset instead of taking care of the mechanized asset itself, the lack of clarity or security target can be practiced.

Blockchain development is an "upset" in the way wherein the information is taken care of and trades occur.

2 Challenges in the Existing Land Registry System

2.1 The Increasing Number of Fraud Cases

There have been a few instances of fakers acting like the dealer of a property. On the off chance that a sham effectively pretends to be a landowner, they may get everything after fulfillment and getaway with the funds. In many cases, the two merchants and purchasers were uninformed of the misrepresentation until found by the land registry as a major aspect of a spot check workout.

2.2 The Involvement of Middlemen and Brokers

Intermediaries and agents are an indispensable piece of each huge business as they find out about market contributions. Purchasers and sellers, as a rule, like to call them to construct a full help team. As an outcome, purchasers obtain a more profound comprehension of the market and recognize lower/more significant expenses for the exchange. Go between accumulating required data from dealers, distinguishing blunders, and deciphering and encouraging the usage of land transactions. Since land is a huge business, it includes countless players, including representatives, moneylenders, mediators, and nearby governments. It prompts extra costs, making the whole biological system costly.

2.3 Human Error/Intervention

As of now, updates to the land registry records are made physically, and the exactness of those progressions relies upon a specific person. It implies that the land registry is progressively powerless against human errors. Human intercession can build the odds of blunders in the land registry framework.

2.4 Time Delays

The current land registry expends a ton of time to finish title enrollments. There could be a hole during finishing and registration. Many legitimate issues can likewise emerge during this long hole. For instance, consider the possibility that you need to serve the landlord's notification to break a rent where the property has been sold. Such issues can make the whole procedure postponed, and purchasers need to hang tight for quite a while.

Table 1 Blockchain versus traditional databases

Properties	Blockchain	Traditional databases
Immutability	High	Medium
Peer-to-peer transactions	High	Low
Traceability of transactions	High	Low
Verifiability of transactions	High	Low
Data transparency	High	Low
Security	High	Low
Latency and transaction speed	Low	High
Scalability	Low	High

The blockchain is compared with the traditional databases with various properties, and the findings are given in Table 1.

3 Literature-Related Land Registration System

Mattila [1] gives a diagram on blockchain advancements as a study paper. In electronic systems, data is regularly transmitted by copying it, beginning with one spot then onto the following. One of the key issues right presently is the means by which to watch that the information got from the framework is legitimate and bleeding edge. While this is unquestionably not a particularly irksome issue to unwind in itself, so far all the courses of action have required placing trust in someone. All things considered, taking the statement of an accepted master for the validity of data is eminently fine, but at this point and again, it is really not. Also, regardless, when it is fine, consistently depending on a center individual is normally exorbitant, and it would be altogether more affordable if we essentially didn't have to. The wording around the whole marvel is still energetically moving. Caught in it everything, it will, in general, be difficult to shape an undeniable picture on blockchain development and the miracle that includes it. In view of all the exposure and vitality, the headway of the blockchain organic framework is normally observed to progress so rapidly that in order to keep up, there is as often as possible an inclination to endeavor to make a dive too significant too quickly.

Blockchain is an advancement section essential to the computerized cash as a movement of data prevents that are cryptographically attached together. Using this chronicled foundation, cautiously talking blockchain is only a basic data structure with passed on multivariant concurrence control. Right when the term at first got notable, regardless, there was only a solitary valuable use of this data structure in nearness: the Bitcoin computerized cash. On account of the nonappearance of an earlier building, Bitcoin, by need, contained the entire advancement heap of the passed on accord design. At the moment that a database ought to be modified by various social occasions at the same time in a covering way, either their progressions must be somehow consolidated or different contrastingly balanced adjustments of a comparative database will rise. All things considered, the individuals from the framework are acting really—or more to the point—that a large portion of the entire framework isn't plotting against the others in a composed manner.

3.1 Blockchain for Recordkeeping

Lemieux [2] gives an archival theoretic evaluation of blockchain. Governments and associations around the globe are starting to take a gander at the use of this innovation, and some have just executed it. The primary attracting card of this creative innovation is the creation of unchanging reliable records without the requirement for a confided outsider. A legitimate model is the proposal of land. In traditional land moves, the system routinely begins with the posting of the property on a land advert, the exchanging of understandings during dealings over the expense, and the satisfaction of the arrangement by enlistment with a state-run land titles selection authority.

The issue with the standard method to manage land moves is that, in any occasion in specific districts, the system is moderate and cumbersome, being routinely reliant on manual records of trades by means of land enlistment pros, and open to coercion and pollution. Different domains are investigating various roads with respect to the utilization of blockchain advancement to address these issues. Chronicled science is stressed over the drawn-out protection of substantial records, which is consistently misidentified as being connected particularly with storage facilities of old, much of the time uncommonly dusty, tomes.

The old hypotheses and norms of archived science are up until now noteworthy today, and they apply as quickly to automated records and recordkeeping in regards to the dusty old volumes of a long time ago. These speculations and norms began to be sorted out in the medieval occasions with the primary school course in the precursor to contemporary recorded science, a course in notarial articulations, offered in 1158 at the University of Bologna. In the seventeenth century, the correct examination of records (called diplomatics) got out of a need to set up the authenticity of medieval reports when there was an extending number of misrepresentations related to European legal conflicts. As chronicled files of crude believability were much of the time presented as confirmation of rights, "the prerequisite for elective techniques for setting up validity extended, and methodology of story examination began to be made and formalized." These methodologies were first organized in 1681 by Dom Jean Mabillon in De re Diplomatica Libre VI, which recalled direction for the affiliation and movement of record work environments, including personnel, rules and the strategy of record creation, coordinating, amassing, and preservation. The instruction of diplomatics spread to assets of law across Europe, and in 1821 incited the foundation of the Ecole des Chartes in Paris. Starting at now, chronicled guidance stretched out to recall the examination of records for the assistance of both legal and legitimate research, thus setting up the structure for contemporary reported science.

Additionally, as the kind of records has been created through time, from cuneiform on earth tablets to papyrus, to wax chambers, paper, and now electronic systems, the point of convergence of reported science has moved from these old physical sorts of records to additional modern propelled structures. Reported science directly has an inevitable perspective in its accentuation on the arrangement of systems (broadly portrayed as including human and particular establishments and techniques) that achieve the drawn-out assurance of genuine records, a synchronous edge in the dynamic shielding of the authenticity of records for an astonishing length cycle, much of the time in chronicled foundations, and a survey point of view in the assessment of the legitimacy or validity of records, which is normally insinuated as cutting edge records wrongdoing scene examination.

Relentless quality is one a greater amount of the essential segments of unwavering quality from an archived science perspective. In authentic science, the term steady quality insinuates "the unwavering quality of a record as a declaration of fact; that is, to its ability to speak to the real factors it is about." Records have had to affect some genuine showing and to memorialize such acts. As needs are, a reliable record will fill in as an impression of the real factors about the showings it arranges and be as such an "extraordinary" depiction of these exhibits and the real factors identifying with them. A record can sub for the exhibit itself.

Subsequently, an exceptional copy of a land title enrollment speaks to the trading of title to a land bundle heavily influenced by another landholder. To achieve reliability, records must have three characteristics: summit at the motivation behind creation, consistency with formal standards of creation, and "intuitive nature." In reported terms, the zenith is associated with the worth-based nature of records and insinuates the proximity of the impressive number of segments required by the creator and an authentic legitimate structure for the record to be prepared for making results. This customarily fuses checks and dates of creation. To appear, an understanding accessible to be bought of land that doesn't have an imprint and date would not be seen as complete. Summit as a reported science thought is thus normal for the record and associated with its customary properties.

A solid record is furthermore one that has physical and formal parts, which are consistent with legitimate records of similar provenance (e.g., whether or not the ink used to make a report is contemporaneous with the document's demonstrated date, or whether the style and language of the record is unsurprising with other related chronicles that are recognized as veritable). Finally, dependable records will have intuitive nature. This implies the way that, normally, records are made over the range of business or step by step life and are thus not commonly arranged intentionally to scatter data or evaluation, as, for example, books or various disseminations.

Taking everything into account, they have generally been thought to have attributes of unselfconsciousness that help their steady quality as records. This idea bolsters the genuine "business records uncommon case to drivel" rule, which recognizes a record as speaking to the real factors suggested in it by the greatness of the intuitive idea of its creation. Understanding the speculation and principles of fundamental solid recordkeeping as clarified in archival science gives an accommodating structure to the evaluation of blockchain-based recordkeeping systems that show to give trusted, perpetual records. Using a chronicled theoretical evaluation structure, it is possible to recognize gaps in, or perils as well, the exactness, constancy, and long stretch validity of such systems. Understanding these deficiencies can coordinate the way toward plan improvements that address openings in this innovative new setup of advances. If not kept an eye on, such gaps could thwart the viable assignment of blockchain-based recordkeeping courses of action. Imaginative work to character recordkeeping structure decisions and trade-offs will lead, as time goes on, to better particular, and downstream, social outcomes.

In [3], blockchain and smart contracts are examined. This paper shows a prologue to the present cutting edge of the blockchain and smart contract types of progress. Blockchain is a lively dubious advancement changing into a key instrument in the normal economy. The blockchain-based smart contract expects to, in this manner, safely execute the important commitments of an understanding without the help of a solidified execution authority. The smart contract runs over the blockchain to invigorate, execute, and support a perception between unbelieved parties without the impedance of an outsider to trust in it as this smart contract is an executable code that runs as per some set conventions on the blockchain. Sharp contracts have two or three highlights that serve the objectives of social worth and goodness. The paper shows the fundamental basic data about the structures of the blockchain and smart agreement movements and provides an evaluation of the various strategies for thinking utilized in the smart contracts. The issues looked inside the smart contract progression are diagrammed.

The four key issues are seen as requesting, security, protection, and execution. We review case events of the use of the blockchain in different business parcels like land, tossing a democratic structure, and creation compose. The paper means to help a master with taking care of the 10,000 foot point of view on the blockchain improvement and also to help the choice methodology of reasonableness of the progression to an express application zone. There are different applications where smart contracts can be applied to. A touch of these applications as of late is utilized in a snare of thing and awesome property, music rights of the board, and e-trade. Regardless, the Internet of things and the canny property are shared between different individuals rapidly through the web.

Internet of things (IoT) uses instances of blockchain-based smart contracts that permit indisputable focus to locate a decent pace warily without the impedance of a distance. Second, the music rights of the directors are a potential use case to record the possession advantages of music in the blockchain. Astonishing contract plays a gigantic standard in music associations as any music will be utilized in any clarification, the proprietor will be redressed, and this construes that smart contract will endorse the business relationship to pay the proprietor with the advantages. Third, e-trade is a potential use case to empower the exchange between untrusted parties; these social affairs are the merchant and purchaser without a confidence in unapproachability. The keen contract is an executable code that bolsters, executes, and actualizes an understanding between untrusted parties. Ethereum is by and large saw wide adoption of blockchain arrange for settling on smart understandings. There are up until now different specific openings looked at during the execution of the smart contract that should be tended to in future assessments. A piece of these holes is perceived in the creation with some proposed game plans. We aggregated these issues into four classes, to be unequivocal, arranging, security, protection, and execution issues. There is comparatively a nonappearance of spotlights on the adaptability and execution issues of the smart contracts. Plus, there is a nonappearance of spotlights on passing on smart contracts on various blockchain courses of action other than Ethereum to see bad behaviors in smart contracts.

3.2 Blockchain in Land Registry Systems

In [4], with regard to land organization, blockchain innovation has the accompanying applications: (a) title deed enlistment, (b) time-stepped exchanges, (c) multiparty straightforward administration apparatuses, (d) carefully designed account framework, (e) catastrophe recuperation framework, and (f) compensation and remuneration in poststruggle zones. The enthusiasm for blockchain has spread to the land organization segment, prodded on partially following news in May of 2015 that a Texas-based organization, Factom, would assemble a land registry framework dependent on blockchain innovation in Honduras; however, these cases have as of late been demonstrated to be distorted or potentially misrepresented. Factom is a long way from the main organization attempting to apply blockchain to the land division nonetheless, with gatherings, for example, Bitland in Ghana, ProSoft Alliance out of Ukraine, and Ubiquity from the United States, all professing to have incorporated parts of the innovation into land framework contributions. This paper dives further into how precisely every one of these associations is joining blockchain and what their arrangements are for what's to come. The idea of a straightforward, decentralized open record could, without much of a stretch, apply to the land data of the executives, where the land registry fills in as a database of all property rights and chronicled exchanges.

The additional advantage of utilizing blockchain innovation is that one can escape from a brought together database, which time and again could be powerless against hacking, abuse by framework directors, or even normal or artificial fiascos obliterating the server farm. In assessing potential utilizations of blockchain innovation, some potential uses ascend to the bleeding edge, for example, (1) time stepping of exchanges likened to virtual authorization, (2) calamity recuperation as the framework doesn't depend on a solitary server farm, (3) recording of subtleties in a sealed and changeless condition, and (4) utilizing "colored coins" to oversee registry subtleties.

Colored coins are utilized to speak to and track unmistakable and impalpable resources utilizing the blockchain. This component of the convention can be applied to land organization by speaking to the responsibility for real estate parcels by a solitary or various tokens. In [5], the metadata appended to the token can be utilized to follow open registry subtleties, for example, size, GPS arrangements, year manufactured, and so forth. Subtleties of possession, for example, liens, or character can be scrambled by the land head, so just those with the right private key can be indicated by the data. Any individual who is associated with the web can freely confirm and follow the responsibility for token utilizing square wayfarer programming.

4   Proposed Model for Using Blockchain for Land Registration

The way toward speaking to the land with a colored coin token is called shrewd property. It can help in the land organization since it can give a simple method to enlist and move a property and help forestall deal extortion. In nations where enrollment of land is troublesome, a minimal effort authentication of possession can be given from a PC. The subtleties attached to the responsibility for token can be put away in the metadata and can be found with a web address. This encourages the land organization's obligation of having something to allude to as systems become increasingly vigorous or in case of a contest.

Private keys might be utilized to sign records or exchanges to guarantee that the individual is the genuine proprietor. Along these lines, the colored coin tokens are a method of forestalling deal misrepresentation. Tokens on blockchain made to speak to the property are currently custom to similar kinds of scripting language and applications talked about in savvy contracts, for example, the closing down of a public accountant, a region assistant or land overseer, and by the proprietor of property take into consideration keen agreements around the exchange of property and deals to be executed. Boundless conditions can be met to decrease extortion and for installments to be sent on schedule. Moneylenders and title organizations can put assets into trustless escrow accounts, possibly to be moved if the installment is gotten or the best possible marks are introduced.

So there is no confirmation of proprietorship because there is no assumption of precision and legitimacy. The main assumption conceivable in a blockchain is an accurate assumption, for example, genuineness. Be that as it may, a portion of the previously mentioned issues could be overwhelmed with a private blockchain, diminishing the excavators to explicit land enlistment centers, setting up another convention in the hubs to control formal and considerable viewpoints. An away from of duty ought to be drawn up if there should be an occurrence of blunder. Moreover, the blockchain ought to have the option to offer access to nonlegally binding acts, for example, legal choices and managerial acts on the grounds that the Bitcoin blockchain configuration is restricted to contracts. In [8], a proof of concept arrangement of land enrollment framework in Thailand is talked about. Land title deeds are noteworthy chronicles that can be used to check ownership and move the history of landscapes. In Thailand, the Branch of Lands works and empowers the trading of ownership and various exercises related to lands. The present development used in the assignments as often as possible encounters issues with the unfaltering nature of exercises and complexities that cause delays. This paper looks at a proof-of-thought (PoC) of the mix of blockchain in the land enlistment structure to improve the trustworthiness of limit and diminishes the methods all the while.

The strategy of advance concurrences with banks is used in the PoC, and an Ethereum-based private blockchain is made and attempted. Execution appraisal of our proposed structure can manage in any occasion 26 trades for each second over an extent of different trades of interest. The results show promising potential both in the abatement of method and the introduction of blockchain for the proposed use case. Blockchain development can help increase straightforwardness and direct trades even more securely. We accordingly carry this favorable position to make applications as a PoC that can be applied for land vault systems. Regardless of the way that there are various requirements to apply the proposed structure in authentic conditions, this paper has developed a system as close to certified data and real techniques as possible, for the favorable position and when in doubt for the relationship to have the alternative to improve later on.

This is an early work on the utilization of blockchain to land selection systems, especially for land selling trade. We lead execution examinations focusing on the throughput in terms of different completed trades executed each second. Results show that our proposed structure can manage on any occasion 26 trades each second over an extent of different trades of interest. Future work will research more understanding of the security parts of the system, the settings of changing number of centers, and more limits in the sharp understanding. Besides, Ethereum was picked for this endeavor due to the pervasiveness and the availability of instructional activities and learning material. For instance, hyperledger fabric might be dynamically suitable for the use case and should be examined as alternatives. Usages of our system for a title deed with various landowners will be inspected in what's to come. Ref. [9] discusses two countries' use examples of land record modernization by grasping blockchain advancement. Through the cases in Honduras and Georgia, the authors review how sociopolitical and specific issues sway the IS readiness of open affiliations while grasping a creating advancement. While the two countries united with private firms to get an aptitude in blockchain, one case was less successful than the other. In Honduras, the nonappearance of an expansion across the nation land vault with generous likewise, complete land records, similarly as political assurance from evolving business, of course, shut down the blockchain adventure. Strikingly, a strong open private relationship with a political buy-in, close by logically present-day and strong mechanized land records, empowered the gathering of blockchain for the land vault in Georgia.

The assessment of these two cases has any kind of effect to recognize engaging and obliging components related to the digitalization of open records, in addition to the choice of land-library blockchain exercises. While these endeavors don't rely upon the improvement of new development, they do require process update and mechanical status. As these two cases show up, the blend of sociopolitical components with advancement-related elements, for instance, structure and accessibility, make the conditions for the accomplishment or frustration of forefront digitalization exercises. The examination of these two relevant examinations helps with recognizing engaging and convincing components related to the digitalization of open records and the choice of land-library blockchain exercises. While these assignments make an effort not to rely upon the advancement of new development, they do require IS readiness. Concerning, the affiliation needs to assess the state of their IT system and update structures. The experience of Honduras furthermore focuses on the hugeness of IS accessibility to move a progression.

The essential system and methodology accept a huge activity in the chance of blockchain digitization of land records. For this circumstance, the creation of a PPP brought the basic imaginative authority and anyway revealed the action to the risks originating from the misalignment of interests between the two social occasions. On the other hand, the Georgia case plots how IS planning is practiced by uniting a legitimate establishment with a powerful association with an advancement firm. The authors acknowledge that information systems research can give the course later on new development and utilization of blockchain-related endeavors in the open part, particularly in making countries. Future work will consolidate additional gatherings to other key accomplices and incorporate additional exercises of using blockchain for land titling in Brazil, Sweden, and Dubai. Ref. [10] shows us the difficulties of utilizing blockchain for land enlistment systems. Blockchain systems are contended to be unsuited for a move of genuine property rights, at any rate, under a precedent-based law framework.

The blockchain development may end up being a helpful vehicle for moving lawful titles to low-esteem resources or resources with a constrained time frame of realistic usability, neither of which has a one-of-a-kind character. Be that as it may, genuine property exchanges don't fit inside this character. The argument is introduced in the following way. Subsequent to presenting what the authors comprehend the operations of a blockchain framework to be, thought is given to its utilization for the exchange of lawful title to both substantial and impalpable resources. This empowers them to more readily outline and understand the issues that at that point develop when we look to use blockchain standards for a move of genuine property rights, bc thcy human or ethereal in nature. The uncertain issues that rise up out of this make it important to set up who conveys the hazard if an exchange turns out badly. The decisions give off an impression of being either the clients or the administrators of the framework.

In the event that a familiar framework has to drastically withdraw from what they comprehend to be key indicia of a blockchain so as to securely work in a dependable way, it is not, at this point, a blockchain framework. On the off chance that such flights are unavoidable (which this chapter holds to be so), we should forsake the descriptor of the framework being "blockchain" in nature. Proceeded with the utilization of this term is misdirecting and will possibly prompt expanded disarray about what we mean when we discuss a blockchain framework. This chapter endeavors to sidestep the "publicity" and manage the hard-hitting real factors of the vital necessities that should be defeated before execution of a registry, genuinely blockchain in nature. This chapter reasons that the blockchain idea (as it is comprehended to work) is unsuited for the exchange of genuine property rights. In any event regarding what is called its "hued coin" application, blockchain neglects to address key issues for genuine property exchanges, chiefly the requirement for free confirmation and control.

On a fundamental level, blockchain is a direct mode for a move of advantages, without outsider intercession. It seems to have worked best in territories where the supporting law strengthens the conviction of exchanges. To the extent that block-chain has worked well, it's realized that triumphs will, in general, have been centered around fungibles, which are resources with no specific imprint or character or low-worth consumables. Given, under the blockchain development, the proprietor of the colored coin is the sole guardian regarding what data is put away, there would likewise seem, by all accounts, to be a constrained chance to draw out into the open significant qualification to bargain issues.

Such concerns sway both the nature of the title and general law limitations influencing the proprietor's capacity to pass on. As in Ref. [6], blockchain's fascination is "attached in a propensity to overestimate the intensity of private requests and to limit that of confided in delegates which frequently prompts disappointed desires." Blockchain defenders seem to disregard a registry's prime capacity (or chance) of giving a component of free checking or confirmation of the provenance of the put-away information. Given, under the blockchain development, the proprietor of the huge coin is the sole guardian regarding what data is put away, there would likewise seem, by all accounts, to be a constrained chance to draw out into the open signifi-cant qualification to bargain issues. Such concerns sway both the nature of the title and general law limitations influencing the proprietor's capacity to pass on. As Arruñada proposes, blockchain's fascination is "attached in a propensity to overes-timate the intensity of private requests and to limit that of confided in delegates which frequently prompts disappointed desires." Blockchain defenders seem to dis-regard a registry's prime capacity (or chance) of giving a component of free check-ing or confirmation of the provenance of the put-away information.

5 Blockchain in Healthcare Industry

Utilizing the existing research and building upon it further, researches as well as industries have both tried to delve deeper into this and find better and enhanced ways to implement blockchain for various aspects in the healthcare industry. Blockchain can have a strong impact on the healthcare industry. In this chapter, we deal with the ways in which blockchain has been implemented and what problems are prevalent in those. We discuss how these problems are tackled in the industry at present and how these can be resolved. We go through some of the applications blockchain may help in, such as patient monitoring, sharing of records, interopera-bility, and secure authentication. Since IoT is emerging as the next big thing, the combined use of IoT and blockchain is a blossoming thing and is driving the soft-ware as well as the hardware industry further at a faster pace. This chapter aims to survey many of the papers that are available on the topic and present a coherent view of all the technological challenges that are yet to be overcome for blockchain to be widely accepted by the healthcare industry.

Blockchain in the healthcare field attempts to solve the rising challenges in the medical field like security, which has become a major issue with respect to the sharing of patient's data securely and difficulty in maintaining health records, so to solve these issues, blockchain has emerged as the best solution as blockchain is a peer-to-peer decentralized network, which means that no single authority can have full control over the system and by blockchain technology we can store all the useful information of the patients in the blocks and they are highly secured as no one can tamper them and no one can modify the information that is stored in the blocks. With integration of new technologies like machine learning, IoT, robotics have given new heights to the medical field as these new technologies have made our life easier.

Nowadays various decentralized mobile apps are coming in the field of healthcare, which uses blockchain to store the patient records in a secure way, and through this app, patients can view their records and their prescriptions, thus making their life easier. Nowadays, with the help of machine learning and blockchain architecture, a new technology has come, which is also known as a remote patient monitoring system. It allows doctors to access information with wireless communications, thus making a patient's life easier. Blockchain in medical fields has done many advancements as blockchain not only provides security but also has resolved many issues in the healthcare fields.

The following are the medical sector applications that blockchain can take care of:

A. Clinical Data Sharing: An essential and key utilization of blockchain in medicinal services is the sharing of clinical information among different elements in the framework. EHRs and EMRs contain profoundly basic and private clinical data identified with the patient, which should be safely put away, shared, and handled.

B. Global Data Sharing: There are also certain occasions where patients travel outside their own country for tourism purposes or for any other reasons in order to provide better health services; respective doctors/hospitals of the other country should have knowledge of the patient healthcare information.

C. Maintaining Medical History: Sometimes, patients visit disconnected hospitals, and thus the overall chain of the medical history might not be available.

D. Research and Clinical Trials: Clinical trials speak to another key and significant procedure in the human services part that requires suitable checking at each phase of the process.

E. Healthcare Data Access Control: In the case of EHR/EMR, various healthcare service providers are associated, and patients are not fully aware of the parties that are accessing, storage, and sharing of their medical data.

F. Drug Supply Chain Management: Medication supply management is vital in the cutting-edge medication industry; however, it, despite everything, experiences different complexities and misfortunes considering counterfeiters and pilfering.

G. Billing/Payers: Traditional modes of patient billing systems are very complex and are exposed to billing-related frauds.

5.1 Usage of Blockchain in Monitoring Health Through Wireless Devices in Remotely Access Area

Advancement in biomedical and healthcare is always considered a major issue that has to be taken care of. The current generation business concern about personal health until they come across a major health issue such a mishap can be always blamed due to a busy and stressed lifestyle of living.

Therefore, we need an architecture that can measure and detect any abnormality in the health of a human, a system that could predict disease from the symptoms and can be consulted to a professional before any critical situation takes place. The ongoing developments in the field of Internet of things (IoT) would help in exponentially increasing the rate and strength of connection between the wireless devices and the user with the help of faster and easier data transfer. Recently, with the arrival of blockchain, the entire scene of access, transaction, and storage management has completely changed and in no time will be taken over by the technology.

An alternative approach for monitoring and providing help with the help of telemedicine technology to the remotely access area profile connection is not available. The architecture developed for the particular situation consisted of pulse oximeter PCG mg and important tools for diagnosing a user. As the data can be easily stored in the form of small packets in the local server and later transferred to the main server, the following system can work both online and offline. The blockchain architecture used for the system can help hide the personal identity of the person and provide authentic and accurate data of the person to the assigned doctor.

The information is stored in two types of blockchain:

- Personal healthcare blockchain
- External record management blockchain

5.2 Usage of Blockchain to Create a Tamper-Resistant Database

The following data of patients is collected with the help of blockchain. The following application records daily report with the help of wireless devices like smartphones and smartwatches. The following application only focuses on wireless devices like smartphones and the architecture of network servers. Open-source blockchain platform hyperledger fabric version is used to operate the following system because the following jobs in the platform is widely used and is open source. The study conducted by the following research papers consists of four validating peers. The main function of friendship songs of the membership service was authentication for wireless devices like smartphones. The validating peers had the duty to perform the main functions of the blockchain platform. Out of the four VPs, one became the leader of the architecture and can provide authentication to the CBTI clients. When the request is accepted by the leader VP, the information is sent to all other VPs. The first transaction from the CBTI is sent to the leader VP under after each day after the other VP carry on the chain and the initializing process. The VPs that initialize the transaction for the chain return a hash value from the executed result. The consensus algorithm famously called the practical byzantine fault tolerance algorithm is followed. The following transaction created based on the hash value generated is called a block. This block contains all the information about the previous blocks in the form of hash values and the hash value of the current block. The shown upcoming blocks will contain information about the previous block, which is called blockchain.

The following application was tested, which consisted of four validating peers and one membership service. The application was initialized to collect user data for 2 days and create each of the validating peers. The details of steps were login into the CBTI system with a login ID and password after which the initialization of data capturing for the patient takes place. The Sachin code is developed consisting of information in the format of JSON and is added blockwise. It was ensured that a block was generated, and the connecting block contains the information about the previous block in the form of hash. The current hash value of the block was matched with the previous hash value, and the information was compared conforming to the working of the system.

6   Risks and Vulnerabilities in the Blockchain and Cryptocurrency

In Ref. [7], the risk of security when using cryptocurrencies is discussed. In cases like shopping online, it stated that the third-party trackers, which get the data from the company to advertise, can use the information of the purchase and end up identifying the unique address of the user and finally their identity. Also in another attack, it explains us about how two different purchases done by the same user can be linked to find personal information about the user. However, it also told us how the data is leaked and how we can identify that leakage using web crawlers. After that, they also analyzed data on how much transaction-related information is actually passed onto the third parties. This paper gives us a clear insight into what data is leaked and how that leakage can be mitigated. It also makes us aware of how other parties can track and misuse our not-so-relevant information to find more about us.

In Ref. [8], the fraud risk assessment that can happen during the blockchain transactions is stated. The author has talked about the double-spending attack as a fraud, which is something that has been solved. Double-spending attack is where a person makes two transactions, maybe one sending the money to himself and another to a person he might want to send it to. In such cases, the person ends up spending the same Bitcoin twice, but now that is not possible as we have mechanisms that can be used to stop double-spending attack. Here, we take into consideration the private and public key, which is unique for every Bitcoin. The author has also calculated the PDF and SF of the double-spending attacks. Later, the probability of a successful attack is derived. The paper is very insightful and analytical, giving lots of mathematical proofs.

In Ref. [9], the author explains how important Bitcoin technology is nowadays and how it is emerging as the financial backbone, but he also states the question of it being the financial backbone is correct or not keeping in mind the vulnerabilities and risks it has including lots of possible attacks, not too secure.

In Ref. [10], the fast development in the selection of blockchain innovation is stated, and it is reported that the improvement of blockchain-based applications has started to reform the money and monetary administration businesses. Past the exceptionally pitched cryptocurrency Bitcoin, basic blockchain applications go from exclusive systems used to process money-related exchanges or then again protection cases to stages that can issue and exchange value shares and corporate securities. The paper stated a few facts about terrorist financing and money laundering, which helps not only with finding their solutions but also gives us insights into how to improvise existing methods. They have also talked about tax treatment with respect to virtual currency criminal implications, stating that while blockchain exercises are not innately (or even ordinarily) criminal under US law, some law authorization examinations may target customary criminal direct that is encouraged by the use of virtual monetary standards and blockchain innovation. In Ref. [11], the software architecture and the corresponding limitations are reported. Being a very detailed paper, this describes a lot about the differences between private and public blockchains. They also talked about another restriction of blockchains in that they are most certainly not reasonable for putting away big data, that is, enormous volumes of data or, on the other hand, high-speed data. This is an innate constraint of blockchains because of the gigantic repetition from the enormous number of preparing hubs holding a full duplicate of the distributed ledger.

In Ref. [12], the criminal behavior and how people are misusing this domain of technology for their selfish needs are discussed. Talking about money laundering, double-spending, and so on, this paper has analyzed the market growth of Bitcoins as well as the criminal cases. Introducing a new term called virtual pickpocketing wherein taking money of other people virtually from wallets is also pickpocketing. It also tells us how the computer hardware, if not proper, can lead to problems as well. Talking about alternative cryptocurrencies like NameCoin, DevCoin, and so on, it is reported that there is a genuine danger of an intermittent unlawful use of cryptocurrency. The extraordinary number of existing markets and the chance of trading effectively Bitcoins by other forms of hard cash make this new technique the ideal vehicle to play out each sort of exchanges identified with tax evasion or unlawful traffic of substances, with all the legitimate ramifications related to the jurisdictional restriction of the criminal demonstrations performed on the Internet.

In Ref. [13], it has been stated that there is a lack of trust and proper regulations are not being implemented. One of the significant dangers presented by cryptographic forms of money is their use in criminal operations as they possibly encourage staying away from specialists' control. Illegal tax avoidance can be effortlessly executed in the framework, either using the mixing wallet administrations or with the assistance of cash trades and e-wallets, which don't require character check. In spite of the fact that it is hard to assess the specific size of the issue, cryptographic forms of money could likewise be used by psychological militant gatherings to rapidly and secretively move cash and making withdrawals through unregulated systems of Bitcoin ATMs. They also stated how Bitcoins are being used for illegal favors or goods on the dark market since Bitcoin is a very secure cryptocurrency and hides your real identity to an extent.

In Ref. [14], the returns and extensively used various visualizations to find out the graphs and patterns of the currency's value and stock value are discussed.

In Ref. [15], it has been stated that trades are constantly uncovered from a danger of hacking because users get to them via an online server. The trades use the P2P organization framework to consolidate numbers and letters to users and create private keys with hash esteem. You can think about the trade as a bank. The data of Bitcoin is put away in the Bitcoin center of the trade. In the event that the wallet itself is put on the trade's server, numerous users fear digital assaults, so they move the data of the wallet to their mobile phones, PCs, or other data stockpiles. This is because the aggressor controls the exchanges and takes them to their account while hacking if the data is on the server. Cryptocurrency essentially uses the PoW or PoS strategy referenced above, yet the security administration gave by each trade is extraordinary. They have also talked about the loss of cash from wallets due to poor management because attackers know how to use that opportunity to send Bitcoins from the user's wallet to their own wallets.

In Ref. [16], it is reported that a cryptocurrency system is helpless at a few levels. A portion of these vulnerabilities are hypothetical, yet many have in truth been abused practically speaking. At the individual level, an individual's private cryptographic key can be "taken." If it is put away electronically on his/her PC or cell phone, this "burglary" or hack can be accomplished using malicious email connections or applications or by using keystroke logging gadgets or programming to follow the private cryptographic key as it is composed. Regardless of whether the private cryptographic key isn't put away electronically yet disconnected, for instance using an alleged paper wallet, access to the private cryptographic key will even now permit a thief to grab one's Bitcoins. Also, on account of Bitcoins and different digital currencies, running off blockchain innovation, the most fragile connection will regularly be the end users (counting cryptocurrency trades) as opposed to the uprightness of their ledgers, which seem to remain generally secure.

In Ref. [17], it is mentioned that smart contracts in Ethereum are very prone to being exploited, and this research study elaborates on the security analysis methods on it. Smart contracts are very independent and monopolous agents in critical independent apps and are very important since they have a big capacity of cryptocurrency to perform trusted transactions. Lots of money every year is spent and invested into these types of transactions, and thus work assets and a huge part of it rely on this mechanism. This paper also emphasizes the points why security analysis methods are necessary to make sure that these transactions don't suffer because of the vulnerabilities and stay away from attacks and hacks. It surveys 16 vulnerabilities and takes care of what to focus on for each vulnerability and how to avoid frauds or scams while working for the same.

In Ref. [18], the authors have an elaborate plan for ethereal cryptocurrency to prevent it from under the hood Denial of Service attacks. They have come up with an adaptive gas system in order to maintain the purpose of securing Ethereum-based blockchain systems and in order to secure transactions. Ethereum is a programmable cryptocurrency and has thus many implementations, but in order to secure those transactions, such steps and initiatives are to be taken. It involves an Ethereum Virtual Machine, which is expected to terminate eventually. Failure at regulating gas costs of EVM operations lets hackers and scammers launch DOS attacks on Ethereum. This paper proposes a very wise way to deal with this situation and has also elaborated further applications as well.

In Ref. [19], they present eclipse attacks on Bitcoin and it's peer-to-peer network, which opens a lot of possibilities for us. They review Bitcoin's peer-to-peer mechanism in detail and use that information in their paper to show how they can use that data in their probability assessment of the eclipse attack, at the same time quantifying the resources used in those eclipse attacks. This way, they present their elaborate plan on how they plan to take the eclipse attack on peer to peer and raise the bar as well. In an era of disruptive technologies, such initiatives are very welcome and promoted. It's a great idea to test and even develop more on such analysis, and the outcome of such a study should be used for further study and research in the same domain.

In Ref. [20], it is stated that permissionless blockchains are very vulnerable and the network layer is especially prone to higher risks. Such a system has reached a consensus decentralized, and thus the authors have presented their plans on how they plan to tackle network layer vulnerabilities when it comes to permissionless blockchain systems. They have five requirements basically, performance, anonymity, resistance to DOS, and hiding topology. Apart from this, another measure is the low cost of participation. The design space is also studied in detail, and these two aspects are pointed out as the main focus areas for vulnerabilities and loopholes. Thus, such a model is prepared wherein both loopholes are optimally minimized and thus minimizing the risk.

In Ref. [21], it is mentioned that with the dawn of technology, the power of blockchain systems has drawn a lot of attention as well as concern toward certain aspects of distributed blockchain and the data that it involves. The data is very crucial because there is very much of it and also it's sensitive data. A lot of times, what happens is that scammers use this loophole to steal information that might lead to loss of time, money, and human effort. They propose a plan wherein modern blockchain techniques can be employed to enhance the security of the power grid. There are also simulated experiments that also show the efficiency of the proposed system and thus take care of a lot of loose ends that exist with fancy technology today. This very well takes out the big problem we as users and consumers are facing and the miners and people who work in these industries face to the general public and address severe issues with what is wrong, what can be done right, and what steps we can take to make sure that it is done the right way.

In Ref. [22], it is stated that although Bitcoin is generally assumed to maintain anonymity, a person's anonymity can be played around with if it's linked to other similar transactions. Thus, this paper presents concerns and ways on how to maintain anonymity, which is such a great value proposition when it comes to Bitcoin or cryptocurrency. Untrusted third-party signatures issue random anonymous vouchers and services to people. Blind signatures and Bitcoin contracts are a great combo to deal with this issue and maintain the sovereignty and beauty of Bitcoin. This system plans on fighting the vulnerability of Bitcoin in a very ethical and fair way and also be legally and tactically correct. This system is also very damage proof and with very narrow loopholes.

In Ref. [23], Stated that FAW attacks on Bitcoin systems is a major problem today. A lot of miners today organize group pools where they together mine one block and divide the gains among themselves. But several and certain vulnerabilities affect the safety while people participate in these pools and these vulnerabilities can sometimes be fatal to the entire system. Thus, the authors have presented an elaborate plan to tackle this situation. They talk about a new attack called FAW attacks. The reward for a FAW attacker is always greater than or equal to a BWH attacker, and hence this fact establishes a small system where there are attackers but also guardians in our Bitcoin systems. This results in expectations to see FAW attacks in pools where mining is happening.

In Ref. [24], both game theory and how the principles of game theory apply in blockchain-based edge networks that are secure and reliable are discussed. They consider a blockchain-based model with M edges. The interactions between a blockchain and mobile device and an edge server are blockchain security games where the mobile device sends a request to the server requesting real-time service and sensitive data. Thus, the authors have presented an elaborate plan to tackle this situation. This system plans on fighting the vulnerability of Bitcoin in a very ethical and fair way and also be legally and tactically correct. The equilibria of each game are determined and obtained, and the two NEs exist to release how the punishment scheme impacts the adversary behaviors of the mobile devices that exist today. The way that game theory and its principles have been used here to decode the math that goes into securing such EDGE systems is really appreciable, and this can also further lead to more research on the same domain.

In Ref. [25], it is said that the paper is an empirical investigation on DOS attacks on various domains of Bitcoin ecosystems and entire functioning programs into place. They have also supplied big amount of stats and data from various incidences where such cases have happened and conducted detailed research on the entire fiasco. The conclusion drawn from the research has been categorized and segregated to properly make sense out of it and push the entire idea toward an intention of solving the problem. This very well takes out the big problem we as users and consumers are facing and the miners and people who work in these industries face to the general public and address severe issues with what is wrong, what can be done right, and what steps can we take to make sure that it is done the right way. The use of case studies and actual factual data also strengthens the roots of their claims and logically provides an answer to a lot of problems this upcoming technology is faced with.

In Ref. [26], they studied and did an empirical analysis of a very recent campaign that was labeled spam, which eventually led to a DOS attack on Bitcoin. They developed a method to detect such transactions that fall under the spam category. The entire loss was heavy since it impacted an entire total system and thus called in for a such an initiative.

7 Conclusion

Blockchain is another and energizing innovation that makes an agreeable and synergistic condition wherein all data, information, and pictures from a reviewing task can be accumulated in a solid, certain, and changeless way. With the changing landscape of looking over apparatuses and instruments turning out to be progressively innovative and broad being used, receptiveness to cooperation and new thoughts will increment over the business. This force could be utilized to bring the utilization of blockchain innovation to the frontline. Blockchain-based land libraries will give an immense improvement over the present paper-loaded and now and then awkward advanced procedures. Eventually, it will bolster and reinforce land administration arrangements and systems around the world. However, the complex legality of land registrations has kept this idea just that. Blockchains are anonymous and trustless, and the "code is law"; in something so humane and complex where first come first serve would be more of tyranny than innovation, the land registration system as a whole may not be able to revolutionize land registry as many experts had envisioned it to.

Digital money secrecy is another examination theme; however, it sits at the crossing point of mysterious correspondence and information anonymization, both entrenched fields. Sadly, it appears to acquire the most exceedingly awful of these two universes. Like information anonymization (and not at all like mysterious correspondence), touchy information must be freely, and for all time put away, accessible to any enemy, and de-anonymization may happen retroactively. Also, as mysterious correspondence frameworks (and dissimilar to information anonymization), security relies upon unobtrusive cooperations emerging from the conduct of clients and applications. More terrible, reasonable hints of the framework may not be accessible at the hour of structuring and executing the security barriers. Going to safeguards, we see that our first assault abuses the inborn pressure among protection and web-based business, and our subsequent assault abuses the innate strain among security and the open idea of the blockchain. In this way, all moderation techniques accompany trade-offs. The accessible alleviations separate into three classes: self-preservation by clients, methods that traders can utilize, and elective cryptocurrencies or cryptographic money-based installment techniques.

References

1. Mattila J (2016) The blockchain phenomenon–the disruptive potential of distributed consensus architectures. ETLA working papers
2. Lemieux VL (2017) Blockchain and distributed ledgers as trusted recordkeeping systems. In: Future technologies conference (FTC) (Vol. 2017)
3. Mahmoud O, Kopp H, Abdelhamid AT, Kargl F (2018) Applications of smart-contracts: anonymous decentralized insurances with IoT sensors. In: 2018 IEEE international conference on Internet of Things (iThings) and IEEE green computing and communications (GreenCom) and IEEE cyber, physical and social computing (CPSCom) and IEEE smart data (SmartData). IEEE, pp 1507–1512
4. Thakur V, Doja MN, Dwivedi YK, Ahmad T, Khadanga G (2020) Land records on blockchain for implementation of land titling in India. Int J Inf Manag 52:101940
5. Anand A, McKibbin M, Pichel F (2016) Colored coins: Bitcoin, blockchain, and land administration. In: Annual world bank conference on land and poverty, 14 Mar 2016
6. Peiró NN, García EJ (2017) Blockchain and land registration systems. European Property Law Journal 6(3):296–320
7. Agrawal R, Chatterjee JM, Kumar A, Rathore PS (2020) Blockchain technology and the Internet of Things: challenges and applications in bitcoin and security. Apple Academic Press. https://books.google.co.in/books?id=FCoMEAAAQBAJ
8. Lindman J, Tuunainen VK, Rossi M. Opportunities and risks of blockchain technologies–a research agenda
9. Goldfeder S, Kalodner H, Reisman D, Narayanan A (2018) When the cookie meets the blockchain: privacy risks of web payments via cryptocurrencies. Proc Priv Enhancing Technol 2018(4):179–199
10. Goffard PO (2019) Fraud risk assessment within blockchain transactions. Adv Appl Probab 51(2):443–467
11. Walch A (2015) The bitcoin blockchain as financial market infrastructure: a consideration of operational risk. NYUJ Legis & Pub Pol'y 18:837
12. Yeoh, P. (2017), Regulatory issues in blockchain technology, Journal of Financial Regulation and Compliance, Vol. 25 No. 2, pp. 196-208
13. Rennock M, Cohn A, Butcher JR (2018) Blockchain technology and regulatory investigations. Practical Law Litigation:35–44
14. Staples M, Chen S, Falamaki S, Ponomarev A, Rimba P, Tran AB, Weber I, Xu X, Zhu J (2017) Risks and opportunities for systems using blockchain and smart contracts. Data 61. CSIRO), Sydney
15. Nica O, Piotrowska K, Schenk-Hoppé KR (2017) Cryptocurrencies: economic benefits and risks. University of Manchester, FinTech working paper (2)
16. Liu Y, Tsyvinski A (2018) Risks and returns of cryptocurrency. National Bureau of Economic Research, Cambridge, MA
17. Kim CY, Lee K (2018) Risk management to cryptocurrency exchange and investors guidelines to prevent potential threats. In: 2018 international conference on platform technology and service (PlatCon). IEEE, pp 1–6
18. Low KF, Teo E (2018) Legal risks of owning cryptocurrencies. In: Handbook of blockchain, digital finance, and inclusion, vol 1. Academic Press, London, pp 225–247
19. Praitheeshan P, Pan L, Yu J, Liu J, Doss R (2019) Security analysis methods on Ethereum smart contract vulnerabilities: a survey. arXiv preprint arXiv:1908.08605
20. Chen T, Li X, Wang Y, Chen J, Li Z, Luo X, Au MH, Zhang X (2017) An adaptive gas cost mechanism for ethereum to defend against under-priced dos attacks. In: International conference on information security practice and experience. Springer, Cham, pp 3–24
21. Heilman E, Kendler A, Zohar A, Goldberg S (2015) Eclipse attacks on bitcoin's peer- to-peer network. In: 24th {USENIX} Security Symposium ({USENIX} Security 15), pp 129–144

22. Neudecker T, Hartenstein H (2018) Network layer aspects of permissionless blockchains. IEEE Commun Surv Tutorials 21(1):838–857
23. Liang G, Weller SR, Luo F, Zhao J, Dong ZY (2018) Distributed blockchain-based data protection framework for modern power systems against cyber attacks. IEEE Trans Smart Grid 10(3):3162–3173
24. Heilman E, Baldimtsi F, Goldberg S (2016) Blindly signed contracts: anonymous on- blockchain and off-blockchain bitcoin transactions. In: International conference on financial cryptography and data security. Springer, Berlin, Heidelberg, pp 43–60
25. Kwon Y, Kim D, Son Y, Vasserman E, Kim Y (2017) Be selfish and avoid dilemmas: fork after withholding (faw) attacks on bitcoin. In: Proceedings of the 2017 ACM SIGSAC conference on computer and communications security, pp 195–209
26. Xu D, Xiao L, Sun L, Lei M (2017) Game theoretic study on blockchain based secure edge networks. In: 2017 IEEE/CIC international conference on communications in China (ICCC). IEEE, pp 1–5

Intelligent Traffic Management with Prioritized Scheduling System

S. A. K. Jainulabudeen, D. Sundeep, and S. Rahul

1 Field of Invention

Manual traffic handling is a huge, tedious, and risk-prone job. Traditionally, traffic signals are employed in the assistance of controlling the flow of traffic. These signals are either operated manually or programmed to update themselves after a fixed amount of time.

Both these methods become pretty inefficient when we take traffic density as a parameter (number of vehicles occupying an area at a given time). Fixed time updating of a signal may cause unwanted vehicle stagnation by allotting signal time for a very less density lane; instead, a dynamic approach based on density is much more feasible and efficient. Our approach is to let signals handle themselves based on various real-time factors such as traffic density and emergency vehicles.

2 Background of the Invention

The Intelligent Transport Systems (ITS) has already been deployed in various developed countries, but the cost of implementation and maintenance of the system is very high. Video capturing of the traffic plays an important role in order to make a better way to handle and manage traffic. These video results may be dependent on

S. A. K. Jainulabudeen
Assistant Professor, Department of Computer Science and Engineering, Panimalar Engineering College, Chennai, India

D. Sundeep (✉) · S. Rahul (✉)
Application Development Analyst, Accenture, Chennai, India
e-mail: sundeep.dayalan@accenture.com; rahul.suresh.kumar@accenture.com

© Springer Nature Switzerland AG 2022
P. Raj et al. (eds.), *Blockchain, Artificial Intelligence, and the Internet of Things*,
EAI/Springer Innovations in Communication and Computing,
https://doi.org/10.1007/978-3-030-77637-4_2

their clarity and quantity. Our module uses the technique of video capturing to capture the density present in the lane and then process it accordingly so that it will trigger the signal dynamically without any storage [6].

Our proposed solution first addresses the problem of cost-efficient and standard systems. We developed a device called STC (Smart Traffic Controller), which can be integrated with existing traffic signals by which traffic lights will be triggered based on the traffic present on the road. Instead of detecting all the vehicles on the road, which deals with the enormous amount of data to process and manage, we will simply write words (Slot 1, Slot 2, Slot 3) horizontally on the road, mark them up with slots, and use OCR along with image processing to scan the letters written horizontally on the road (refer to Fig. 1). The presence of vehicles will hide those horizontal wordings; based on that, we will measure the density of that lane. We separated density into three levels from levels 0–3. Each level will have actions, and according to that, traffic signals will be triggered with the help of the STC (Smart Traffic Controller) device. Emergency vehicles (ambulance, fire engines) get tracked down as they are fitted on with our INS (Intelligent Navigation System) module, where once the vehicle nears the signal, the signal will be automatically changed to green to allow smooth flow of emergency vehicles. This will be done dynamically when INS syncs with STC.

It allows the traffic to flow through the junctions smoothly, and there won't be any congestion caused due to unnecessary wait time's injunctions. The CCTV camera on roads along with the modules fitted within the vehicles allows reducing traffic violations and other crimes related to it. It also detects incoming emergency vehicles and allows it to pass through the junction without waiting, which helps save lives in timely situations. It effectively reduces travel time, travel cost, air pollution, and accident risks.

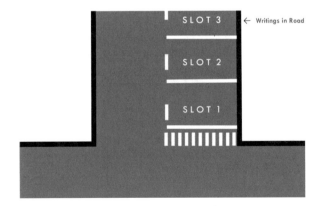

Fig. 1 Base image model

3 Summary of the Invention

3.1 Traffic Density Calculation

The density of traffic will be calculated by detecting the horizontal words present on the road using Tesseract OCR and image processing. This can be done with the help of CCTV present on the roads (If not, need to be installed) [1, 2, 5]. The CCTV data will be sent to the server, and density will be calculated after processing the data. Then, the density value will be sent to the STC device, and it will trigger traffic lights according to that. (Fig. 2).

3.2 Intelligent Navigation System (INS)

We developed a separate navigation system for emergency vehicles by which the smooth flow of emergency vehicles can be achieved. This system provides navigation to the accident location in a single click and simultaneously tracks the location of the ambulance location in the background [4], while moving INS dynamically searches for nearby traffic signals where the STC device will be installed (refer to Fig. 3). When the ambulance enters the radius of the traffic signal, the INS and STC will be automatically synced, and the lane signal will be turned to green by changing all other signals of the lane to red. All these processes will be done independently without any interaction with the ambulance driver, by which no emergency vehicles will have to wait in traffic.

Not only emergency vehicle but also normal vehicle users will have separate INS for their dashboard navigation where the traffic signals are marked in the navigation map along with the lane traffic. The status of the traffic signals will be

Fig. 2 Camera placement over traffic signals

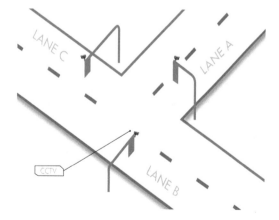

Fig. 3 STC and INS implementation

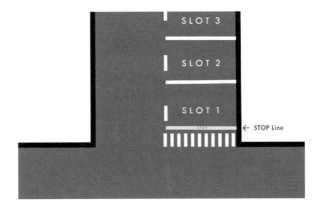

Fig. 4 Traffic violation detection

dynamically updated in this navigation system. By this, the public can reroute based on the traffic, and it will decrease the waiting time in traffics considerably. Even the nearby emergency vehicles will be shown in INS for global users.

3.3 Traffic Violation

The vehicles that violate the traffic rules get detected from the camera; if the signal is red, it scans for the STOP line (refer to Fig. 4), and if any vehicle crosses that line, the vehicle is flagged and the OCR takes the license plate number and stores on to separate data child in the cloud database as flagged. In the case of the absence of CCTV in that road, if the vehicle user utilizes the INS, then INS will throw user details to the server if the user violates the traffic signal.

3.4 Control Interface

This central system, also called the server, will have all the dataset about the traffic lights, users who violated traffic signals, traffic densities of roads, and so on. This cloud database will store all the data and communicate with each STC device dynamically. Also, this interface will help traffic police to control the traffic signals remotely and block particular roads in the case of special events (e.g., VIP convoys). Using this interface, no traffic policeman needs to visit the traffic signal to control or troubleshoot it. Everything can be done remotely. This integrates every single module and serves the entire traffic management system (Fig. 5).

3.5 STC (Smart Traffic Controller)

This is a hardware device that must be installed in existing traffic signals. After that, normal signals will get smarter so that traffic signals will operate automatically based on the accurate traffic present on the road. This device will be connected with servers wirelessly with the help of a private network. The traffic density value will be received from the server, and STC will trigger traffic signals according to that. These traffic signals will automatically fall back to the traditional mode (triggers traffic lights based on fixed timings) of operation in the case of any server issues (lack of internet connectivity). Also, this device can sync with the upcoming INS of emergency vehicles (Fig. 6).

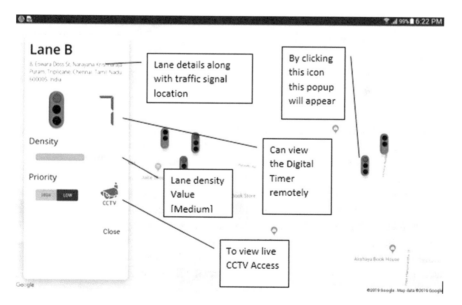

Fig. 5 Control interface UI detailed

Fig. 6 Smart traffic
controller (STC)
installation

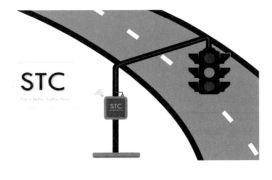

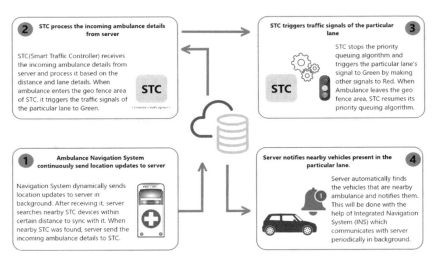

Fig. 7 Emergency vehicle prioritization flow chart

3.6 Emergency Vehicle Prioritization

Ambulance drivers using an INS upon entering a predefined radius from the traffic signals will be automatically synced to the signals they approach so that the signals turn green and provide immediate free passage to the ambulance [3] (Fig. 7).

This is done by tracking the ambulance using the GPS module present in the console of the vehicle. When an ambulance gets near a traffic junction, the traffic light of the specific lane turns green until the ambulance passes. These are updated to a cloud system, from which the information is relayed back to the users in front of the ambulance, alerting them of the incoming ambulance in INS and requesting them to give way for the ambulance. Apart from this, the users are also intimated well in advance about the ambulance's arrival via a voice alert in their mobile phones and INS.

4 Object of the Invention

It allows the traffic to flow through the junctions smoothly, and there won't be any congestion caused due to unnecessary wait times in junctions. The CCTV camera along with the modules fitted within the vehicles allows reducing traffic violations and other crimes related to it (Fig. 8).

It also detects incoming emergency vehicles and allows them to pass through the junction without waiting, which helps save lives in timely situations. It effectively reduces travel time, travel cost, air pollution, and accident risks. By implementing the whole system, no traffic policeman is needed to operate every individual traffic signal. This will effectively decrease the count of traffic policemen standing along the roadside to control the traffic and traffic lights. Global users can dynamically view the traffic signal and density status will be very useful to decide their travel route prior to the departure. Every emergency vehicle driver can operate the vehicles without any interruption by utilizing INS.

5 Working

This system mainly works with the help of the STC (Smart Traffic Controller) device. This device syncs with every other module to make this system fully functional. This STC device communicates with the server with the help of a private network. STC can also work independently when it failed to communicate with the server (offline circumstances). This STC can be installed in any existing traffic lights; thus, it does not need any prerequisites. STC will trigger the traffic lights

Fig. 8 Hardware structure

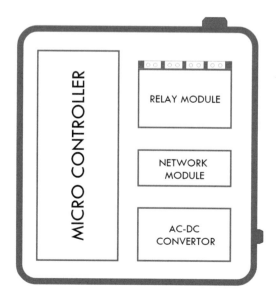

according to the traffic present on the road. This can be achieved by detecting the traffic using the OCR concept. CCTV cameras will be placed in the position, and it continuously detects the traffic density and sends the data to the server. After the data processing is done, it sends the accurate density values to STC, and according to the density values, STC triggers the green lights. All these processes will be done simultaneously without any delays (Fig. 9).

This is how the CCTV cameras will be installed above traffic lights. These CCTV cameras must be installed at considerable heights to get accurate results. The existing surveillance cameras can also be used for this system. The data from CCTV will be stored in the server periodically to allow surveillance backups. The process begins with the data captured from the CCTV cameras at the junctions where it collects all the information like the density of vehicles and on which lane is the high density and sends the data to the server where all the data gets processed and analyzed. The data accumulated will calculate the density and send it to the STC. The STC modules receive the value and change the signal correspondingly. STC will also receive several other values from the server like priorities, troubleshoot instructions, and so on.

To make this system completely utilitarian, it needs a web network. Here comes the most concerning issue of communication between INC, STC, and servers during the absence of internet connectivity, and because of this, the system may glitch. To overcome this issue, we utilized noncellular data transmission, and this system will be utilized automatically during offline conditions. Here the communication will happen through encoded SMS with the help of SMPP (Short Message Peer to Peer) protocol. Once the network is established, the system falls back to normal methodology.

6 The Detection Concept

6.1 Level 0

All the texts (Slots 1, 2, and 3) are visible to the camera, and the processing unit sends the traffic density value as 0 to the STC. When STC receives traffic density as

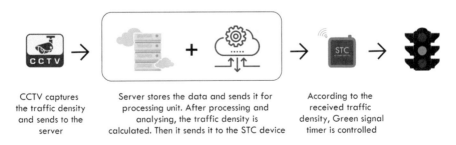

CCTV captures the traffic density and sends to the server

Server stores the data and sends it for processing unit. After processing and analysing, the traffic density is calculated. Then it sends it to the STC device

According to the received traffic density, Green signal timer is controlled

Fig. 9 Communication with server flow

0, STC won't trigger green signal for that particular lane because there are no vehicles, and triggering the green signal will be a waste of time (Fig. 10).

6.2   Level 1

Slot 1 text won't be visible to the camera because it is hidden by vehicles present there. So, the processing unit sends the traffic density value as 1 to the STC. When STC receives traffic density as 1, STC will trigger green signal for 20 seconds (duration can be changed), and it was enough for the vehicles to pass the signal (Fig. 11).

6.3   Level 2

Slot 1 and Slot 2 texts won't be visible to the camera because they are hidden by vehicles present there. So, the processing unit sends the traffic density value as 2 to the STC (Fig. 12).

When STC receives traffic density as 2, STC will trigger green signal for 40 seconds (duration can be changed), and it was enough for the vehicles to pass the signal.

6.4   Level 3

Slot 1, Slot 2, and Slot 3 texts won't be visible to the camera because they are hidden by vehicles present there. So, the processing unit sends the traffic density value

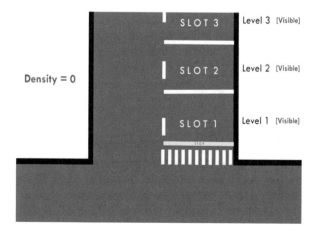

Fig. 10 No density model overview

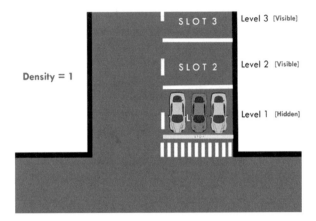

Fig. 11 Low-density model overview

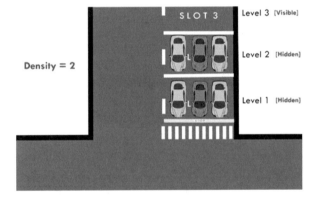

Fig. 12 Medium-density model overview

as 3 to the STC. When STC receives the traffic density as 3, STC will trigger green signal for 60 seconds (duration can be changed), and it was enough for the vehicles to pass the signal (Fig. 13).

7 Conclusion

The project can be implemented in any country with a traffic light with CCTV camera. Cost efficiency is a major factor in the implementation from the installation to the management. Zero Data Storage: We don't store any data for signal prioritization. System Management: We don't have to manage the system as every node is remotely handled and the processing happens at the base junction itself. Prioritization Algorithm: We have developed a unique algorithm for a basis of managing the flow

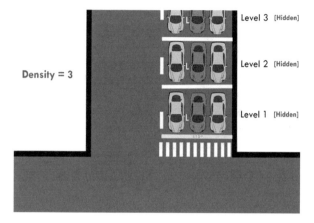

Fig. 13 High-density model overview

of traffic. Minimal Human Intervention: Our idea is to completely move the manual tedious way of human handling. Backup System: Once our system faces any of the issues, the system automatically switches to the backup system having a traditional traffic management system and the issue data will be sent automatically to us so that we can figure out the problem and then remotely fix it if possible. Signal jumping vehicles can be detected and informed to the officials with the CCTV footage.

8 Insights of the Invention

1. Traffic density calculation using OCR having no data storage makes it more feasible in setting up and maintenance in any places. Density calculation based on bay detection instead of detecting every vehicle makes use of less processing and low maintenance of the data.
2. Our controller system STC (Smart Traffic Controller), which we built, is designed to prioritize the signal system with a round-robin fashion and emergency vehicle prioritization. As the system processes the queuing intelligently based on our algorithm, it provides all the lanes some priority and makes to move at the earliest without any manual intervention.
3. Emergency vehicle prioritization is already available in various existing systems, but the systems lag in many aspects. We have introduced a system that overcomes all the hurdles that the current systems face. We have developed our system with high priority for availability, cost efficiency, and standard. The system directs the message to the STC using a distinct cellular network when an ambulance enters the certain entry point marked using the INS, where there is no need for placing a special device in the ambulance to send the signal to the STC. If the system fails to receive the signal, a private network is established so that even if the network loss occurs, the signal can detect the ambulance.

4. Collectively, there is a major problem of detecting the lane in which the ambulance arrives in the existing systems. Since the ambulance is marked on the INS
   (Intelligent Navigation System) in our system, the lane in which the ambulance
   arrives is noted, so the signal passing can be much easier.
5. Our system passes the data to the nearby vehicles from the ambulance through
   the INS. Once the ambulance is in the range with the vehicles, the ambulance can
   instruct the vehicles to move "left" or "right."
6. Traffic violation has been a serious problem in most of the developing countries.
   Our system in the traffic junctions not only maintains the traffic density but also
   detects the vehicles that violate the traffic rules. The vehicles that get detected
   are cross-referenced with the government API data. Violated vehicle users can be
   penalized immediately.

9   Future Direction/Long-Term Focus

The developed system is made on concentrate cloud data optimization and to be
enhanced on the future technologies, thereby avoiding the need for storage of tons
of information in the database and focusing on real-time processing with the help of
blockchain technologies. With the evolvement of enhanced technologies, the future/
scope for this system developed to handle traffic is high, and the minimal cost and
integration with the existing systems make this more efficient and powerful to handle the real-time traffic with ease.

References

1. Shelke M, Malhotra A, Mahalle PN (2019) Fuzzy priority based intelligent traffic congestion control and emergency vehicle management using congestion-aware routing algorithm. J
   Ambient Intell Humaniz Comput. 10(10), https://doi.org/10.1007/s12652-019-01523-8
2. Alsrehin NO, Klaib AF, Magableh A (2019) Intelligent transportation and control systems
   using data mining and machine learning techniques: a comprehensive study. IEEE Access
3. Khan A, Ullah F, Kaleem Z, Rahman S, U, Anwar H, Cho Y-Z (2018) Emergency vehicle
   priority and self-organising traffic control at intersections using internet-of-things platform.
   IEEE Access
4. Yao N, Zhang F (2018) Resolving contentions for intelligent traffic intersections using optimal
   priority assignment and model predictive control. In: IEEE conference on control technology
   and applications (CCTA)
5. Javaid S, Sufian A, Pervaiz S, Tanveer M (2018) Smart traffic management system using Internet
   of Things. In: International conference on advanced communication technology (ICACT)
6. Bommes M, Fazekas A, Volkenhoff T, Oeser M (2016) Video based intelligent transportation
   systems – state of the art and future development. In: 6th transport research arena. Elsevier

Data Mining-Based Metrics for the Systematic Evaluation of Software Project Management Methodologies

Patricia R. Cristaldo, Daniela López De Luise, Lucas La Pietra,
Anabella De Battista, and D. Jude Hemanth

1 Introduction

The management of software projects includes the fusion of science and management. It includes several aspects: direction, scope, stakeholders, risks, planning and control of activities, project requirements, and business objectives. It refers to the project manager's abilities to manage problems related to management and technology. To help, there are numerous project management methodologies and guides on the market. Some of them are PMBOK [1], PRINCE2 [2, 3], APM [4], ISO 21500 [5], SCRUM [6, 7], KANBAN [8], and CRISP-DM [9, 10]. The correct management of projects looks for the conclusion in time and with the desired quality [11]. According to the 2018 CHAOS report, 29% of the projects respect the time, budget, characteristics, and functions required. In contrast, 37% do not respect any of these axes, and 52% of projects experience delays, exceed budget, or implement fewer requirements [12]. This is over 10% of what was reported in 2010. Likewise, the report shows a cancellation of the project without a product of 19%.

P. R. Cristaldo · L. La Pietra · A. De Battista
GIBD – National Technological University, Regional Branch Concepción del Uruguay,
Entre Ríos, Argentina

D. L. De Luise
GIBD – National Technological University, Regional Branch Concepción del Uruguay,
Buenos Aires, Argentina

CI2S LABS, Buenos Aires, Argentina

D. J. Hemanth (✉)
CI2S LABS, Buenos Aires, Argentina

Department of ECE, Karunya Institute of Technology and Sciences, Coimbatore, India
e-mail: judehemanth@karunya.edu

© Springer Nature Switzerland AG 2022
P. Raj et al. (eds.), *Blockchain, Artificial Intelligence, and the Internet of Things*,
EAI/Springer Innovations in Communication and Computing,
https://doi.org/10.1007/978-3-030-77637-4_3

Among the reasons for not reaching the objectives in a timely manner can be cited: insufficient planning [13–15], poor definition of requirements [16–18], lack of skills, problems with the discipline of management, and organization on the part of those in charge of carrying out the projects [19].

Various authors propose the "hybrid" approach, which merges traditional proposals with agile [20–24]. In this methodology, the project managers must focus not only on the final objective but also on the moment when they choose one methodology or another.

These changes in management have been accompanied by numerous efforts to systematically evaluate the quality of management and its effects. Some of these efforts are presented and discussed in this chapter, along with proposals for metrics and indicators to measure and evaluate all aspects and components of project management methodologies and guides. It is thus intended to develop a measurement framework, through the metrics listed, to cross-evaluate different methodologies quantitatively. This allows, among other things, expressing the degree of applicability of the strategies in the different phases of a project and/or in projects in different contexts.

The objective of this chapter is to present and analyze the state of the art in the field of generating evaluation metrics for project management methodologies. A proposal of original metrics and a certain combination of compatible metrics extracted from the bibliography is also made. The sections that follow show the state of the art in metrics and evaluation of management control methodologies (Sect. 2), presentation of a proposal of metrics formulated for managing the scope of a project (Sect. 3), case studies (Sect. 4), and conclusions and future work (Sect. 5).

2   Literature Review

The project management best practice guide (Pmbok) determines ten areas of knowledge present in management: scope, time, quality, costs, risks, human resources, communications, stakeholders, acquisitions, and integration [1]. With this as a basis, the bibliographic search to evaluate the state of the matter is translated, in the present analysis, to the complete management of a project.

The topic related to project management is extremely current and relevant to the sector. To show this fact, an analysis is carried out on the metadata of the indexed scientific publications that contain the following search term: "project management information technologies metrics." This search is limited to the Scopus database of the Library Portal of the National Technological University. Based on the above, a database with 965 publications corresponding to the period from 2011 to 2020 is made. The study presented here uses the bibliometrix library [25] and the VOSviewer software [25] that allows a graphic analysis to be carried out from the generation of maps based on co-authorship, citations, co-citations, and the keyword co-occurrence. Although the analytical work is outside the scope of this work, the initial trend can be observed in Fig. 1, which shows the relevant words of the articles in a cloud-style map. For this, the evaluation by dual count is established, that is, how many times the words appear in the articles and how they are related to each other.

To measure project performance, there are different metrics around the ISO 25010 product/software quality model. In Ref. [26], a generic environment is proposed that is inspired by the principles of Value-Centered Thinking (VFT) and the Meta-Question-Metric (GQM) method to develop performance criteria regarding the values of the stakeholders in the project. GOCAME (Goal-Oriented Context-Aware Measurement and Evaluation) is a metric system [34] that is combined with GQM [33] and C-INCAMI (Contextual-Information Need, Concept Model, Attribute, Metric and Indicator) [35] and associated with the most frequent quality metrics [36–39, 52]. In Ref. [39], a framework based on predictions is developed to measure the behavior of a data warehouse. In Refs. [44, 46], quality metrics are established to evaluate business process models.

Among the metrics, some are developed for specific areas. In banks [27], the possibility of a quantitative analysis for collaborative businesses is presented. The authors apply artificial intelligence with semantics. For SMEs [32], the exploitation of information is measured with metrics that measure different parameters. Three categories are defined: data, models, and projects. In medicine, metrics are used to measure the quality of reconstructed images from CT scans [37]. In wireless communication systems, flexibility metrics are used for mixed numerology in 5G [40].

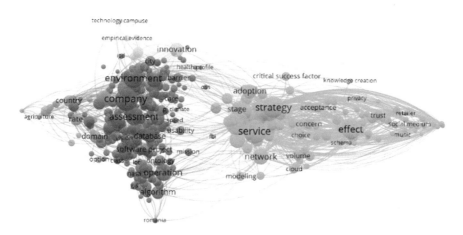

Fig. 1 Cloud map of titles and abstracts As can be seen, there is a high frequency of content around what it does to companies (company), metrics in general (assessment), and, to a lesser degree, the study of critical factors for success (critical success factor), strategies (strategy), modeling (modeling), and services (service). It is interesting that effect aspects (effect) are also reflected. All these indicate factors normally understood in the processes of modeling and control of project management

In natural systems, the success of biodiversity mitigation strategies is measured [40]. In videogames, metrics are established to evaluate interactives with respect to user satisfaction [41]. In cloud computing, metrics are established to measure fluctuations in demand for resources and services [42], scalability [58], and security [43]. In vehicle traffic management, complex network metrics are developed for different urban scenarios [45]. In Ref. [47], a metric approach is presented to store and query data in the management of the Internet of things.

Regarding software metrics, there are several examples. Some allow evaluating the propensity of failures [55, 59] with neural networks [28] and risk assessment [29] with free review and application of indicators to predict the maintainability of the source code by applying machine learning techniques [50, 51, 53] for the readability of the code [57] and to estimate defects in the source code [48, 49, 54]. There are metrics affected only by software design [49] and others at each stage, namely, product, process, test, maintenance, and customer satisfaction, using machine learning algorithms [56].

The metrics presented in the Pmbok [1] evaluate the performance of projects according to the earned value management (EVM) method. However, projects with multiple parallel work paths require special metrics [30]. Other strategies define an index of performance of duration [32]. Specific metrics are also applied in risk management [31, 32]. For each case, to support the project manager, the work of the developers is measured [59]. In [60], agile metrics are defined to monitor and control the best practices of the ISO/IEC/IEEE 12207 and ISO/IEC TR 29110-5-1-2 standards.

This work is aligned to the development of metrics, but not for project management itself, but for the evaluation of the process of applying management models. The works [31, 32] for risk management and those [59, 60] for management in general are taken as a basis. Likewise, when looking for a comprehensive project management metric, new metrics of an original nature are provided based on concepts mainly from NLP.

3   Metrics for Project Scope Management

According to the Guide to Good Practices in Project Management, scope management is the set of processes necessary to ensure that all the work required to successfully complete the project is included [1]. Bearing this in mind, it is important to formalize the initial project document, Business Scope Statement, with a complete description of the list of requirements, in order to lay the solid foundations of a quality project [1]. Despite the fact that this work is part of a comprehensive project to determine comprehensive metrics for the entire process of applying project management models, due to space issues, only those related to project scope management are presented. That is, only the metrics that evaluate how the project management models manage, the scope management aspects, are described. They apply to the case study in Sect. 4. The rest of the model and indicators will be introduced in future publications and are not part of the scope of this work.

3.1 Degree of Business Compressibility (GCN)

According to the literature presented in the previous section, it is important to determine the level of comprehensibility in the Business Statement (mission, vision, scope). To determine a metric from the findings and parameters found, we define:

$$GCN = relPalCla * relParr \tag{1}$$

where

- relPalCla = prom(p/parr)
- p = keywords (nouns <10% or 5–10 words less used)
- parr = paragraphs (number of paragraphs within the Business Statement)

3.2 Degree of Scope Completeness (GCA)

Another important aspect is to calculate the degree of completeness of the scope (eq. 2). The description should contain the business requirements, stakeholder requirements, solution requirements, project requirements, and quality requirements. If one of these elements is not present, it is evaluated at 0. GCA metric is determined as follows:

$$
\begin{aligned}
GCA = {} & pond(RQN) * \log 2(pond(RQN)) + \\
& pond(RQI) * \log 2(pond(RQI)) + \\
& pond(RQS) * \log 2(pond(RQS)) + \\
& pond(RQP) * \log 2(pond(RQP)) + \\
& pond(RQC) * \log 2(pond(RQC))
\end{aligned} \tag{2}
$$

In turn, Business Requirements (RQN) are defined as a set of needs and opportunities of the organization (eq. 3). They must be extracted from the list of requirements. If they do not exist, they are evaluated at 0.

$$RQN = \Sigma(r) / \Sigma(RQN, RQI, RQS, RQP, RQC) \tag{3}$$

- r = count($Z > = 0$)
- RQN ε [0..1]

Stakeholders' Requirements (RQI) are a set of needs of those who participate in the project. They are extracted from the list of requirements. If they do not exist, they are evaluated at 0.

$$RQI = \Sigma(r)/\Sigma(RQN,RQI,RQS,RQP,RQC) \tag{4}$$

- $r = count(Z > = 0)$
- RQN ε [0..1)

The Requirements for Solutions (RQS) constitute the set of characteristics and functionalities of the product or service. They are extracted from the list of requirements. If they do not exist, they are evaluated at 0.

$$RQS = \Sigma(r)/\Sigma(RQN,RQI,RQS,RQP,RQC) \tag{5}$$

- $r = count(Z > = 0)$
- RQNε [0..1)

The Project Requirements (RQP) are the set of actions and processes that the project must provide. They are extracted from the list of requirements. If they do not exist, they are evaluated at 0.

$$RQP = \Sigma(r)/\Sigma(RQN,RQI,RQS,RQP,RQC) \tag{6}$$

- $r = count (Z > = 0)$
- RQNε [0..1)

The Quality Requirements (RQC) are defined as a set of conditions or criteria that the product must satisfy. They are extracted from the list of requirements. If they do not exist, they are evaluated at 0.

$$RQC = \Sigma(r)/\Sigma(RQN,RQI,RQS,RQP,RQC) \tag{7}$$

- $r = count(Z > = 0)$
- RQNε [0..1)

The next section shows how these metrics should be applied.

4 Case Study

The study in this section is based on a small set of organizations in the software industry that work with project management. As can be seen in Fig. 2, the profiles of the respondents are varied (project leader, IT project managers, company owner, and administrative directors without IT training).

At the same time, the type of companies participating in this case study is also varied (see Fig. 3).

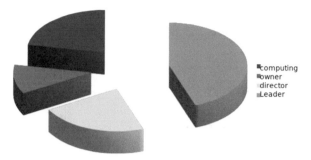

Fig. 2 Respondent profile

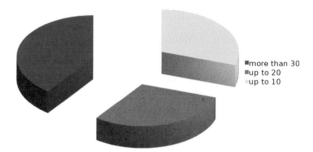

Fig. 3 Company size

The database is divided into two samples: one in which the PMs previously define the scope of the product and the project and the other in which they do not define it. In the latter case, PMs are assisted based on the data requested in the survey. The sample is protocolized through a survey through the LinkedIn social network. Likewise, this section shows the treatment of metrics defined in the previous section, when they are applied post-mortem to the responses obtained, and some discussions about them in the context of project management.

4.1 Demographic Sample

Of the respondents in the sample, the majority are private sector organizations (80%), while only 20% are from the public sector. On the other hand, 40% of the organizations, at the time of the survey, have between 11 and 20 employees, 30% more than 30 employees, and 20% less than 10 employees. Of the validated respondents, 65% are project managers and directors and 35% are project coordinators and members of project teams. Also, 20% of those surveyed certify that they have knowledge of methodologies and guides to good practices in project management.

Another noteworthy fact is that 60% of the respondents confirm that they have defined the scope of the project and the product, but not the other respondents.

4.2 Behavior of Metrics

This section studies the behavior of the metrics designed in Sect. 3 applied to the sample described in the previous point.

As can be seen in Fig. 4, the scopes were not always defined despite the fact that they are all operating companies. The metrics of this study are intended to determine the characteristics of the management documents in relation to the type of company and its core business.

To incorporate and interpret from the subjectivity of the respondent in the scope of real projects, this fact is considered part of the metrics. From these documents and the communicated scopes, Table 1 is obtained, which results from the application of equations 1, 2, 3, 4, 5, 6, and 7 that determine the management of the project scope. It should be remembered that these are part of a set that encompasses

For metrics, it is systematically processed, and nouns, verbs, and so on are labeled. In some cases, some responses had to be reorganized since functional requirements were confused as business requirements, among others.

To complement the above, the linguistic analysis of each scope is performed separately using the Octave NLTK library (c). This allows obtaining numerical complements on the expressiveness and complexity of the descriptive language of the documents and indirectly of the information contained.

Both studies will be related to establishing the adequacy of the metrics proposed to measure this and other factors. Figure 5 shows the behavior of the frequencies of the keywords corresponding to all the responses concatenated to determine the scope.

For better comparison during the discussion of results, those corresponding to the same characteristics have been placed superimposed.

Fig. 4 Defined scope

Table 1 Preliminary results of the application of metrics

	GCN	GCA	RQN	RQI	RQS	RQP	RQC
ID1	0.3	0.01	0.3	0.2	0.25	0.05	0.05
ID2	0.6	0.02	0.13	0.13	0.38	0.38	0.0
ID3	0.25	0.11	0.1	0.1	0.0	0.28	0.0
ID4	0.33	0.26	0.03	0.03	0.35	0.64	0.02
ID5	0.15	0.01	0.15	0.0	0.0	0.01	0.58
ID6	0.35	0.22	0.31	0.32	0.11	0.08	0.12
ID7	0.63	0.22	0.25	0.29	0.12	0.24	0.17
ID8	0.61	0.21	0.68	0.67	0.70	0.16	0.15
ID9	0.68	0.32	0.51	0.33	0.41	0.62	0.72
ID10	0.74	0.31	0.41	0.55	0.52	0.41	0.75

Table 2 shows the comparison between the various characteristics of the text from the linguistic point of view:

- DL: Lexical diversity
- SR: Number of words without repetition
- CR: Number of words with repetition
- PMD: Number of words with length greater than ten characters

In the GCA metric, in turn, the partial values of its components RQN, RQI, RQS, RQP, and RQC are discriminated. Table 3 shows the behavior of the distribution of the words in each ID. The Ls represent a linear behavior, where all the words present identical frequency (value 1). The entries with the value NO correspond to the cases in which the respondent does not answer the question. The rest of the cells present words (one or more), which exceed the average value of the rest and therefore are considered special value for the analysis.

Note that out of 50 entries, 42% present linear entries and 10% are not answered. In total, they add up to 52%.

The next section studies the relationship between these results, the linguistic analysis, and the type of company.

5 Discussion of Results

The scope of the project must describe the characteristics, functions, and requirements of the product or service and its management. This implies that, in the context of this proposal, the defined metrics must be able to establish the degree to which the project scope management satisfies the equations of business requirements (RQN), stakeholder requirements (RQI), requirements of solutions (RQS), project requirements (RQP), and quality requirements (RQC).

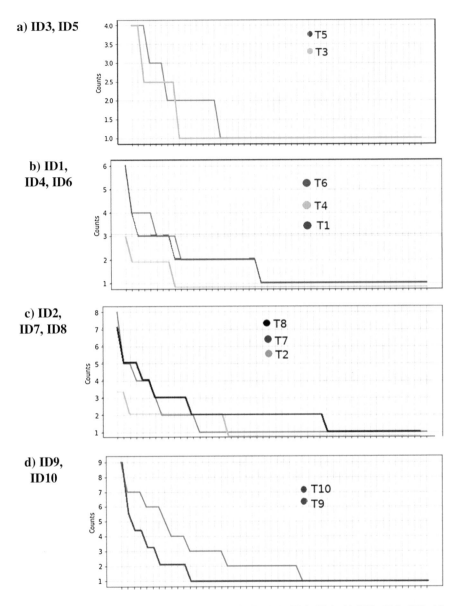

Fig. 5 Word distribution (ID3, 5). (**a**) ID3, ID5. (**b**) ID1, ID4, ID6. (**c**) ID2, ID7, ID8. (**d**) ID 9, ID10

The highest GCN occurs in ID 10, followed by ID 9 and 7 (all greater than or equal to 0.63). This indicates that these companies present a greater degree of the scope of understanding of the business, reflected in its scope. They correspond to companies where the document was defined, but each one has a different size (one large, one medium, and one small). However, in neither case is the owner

Table 2 Lexical analysis of texts

ID	CR	SR	DL	PMD
1	281	129	0.46	37
2	206	113	0.55	37
3	212	121	0.57	24
4	159	99	0.62	18
5	211	141	0.67	19
6	218	109	0.50	37
7	175	90	0.51	30
8	318	153	0.48	29
9	415	210	0.51	29

Table 3 Frequency of distribution of the words in the text per question

ID	RQN	RQI	RQS	RQP	RQC
1	L	L	PAYMENTS	L	L
2	L	KNOWLEDGE	SYSTEM	L	L
3	PAYMENTS	L	NO	INTEGRATION PLATFORM	NO
4	L	NO	HIGH	L	L
5	L	NO	NO	NO	SHOULD BE
6	L	TECHNIQUES	L	L	L
7	AVAILABILITY	L	L	DRAFT BILL PROCESS	L
8	SYSTEM	SYSTEM	MODULE LOAN	PROCESS	L
9	MANAGEMENT	L	MODULE	BANK PROCESS	DATA MUST BE
10	SURVEYS RESULTS	GOVERNMENT SERVERS	APPLICATION SHOULD BE	COMPLIANCE RULES	CREDITS MODULE LOAN

responding but rather personnel specialized in software development management or senior management of projects at scale.

Note that the general distribution of words (Fig. 5) shows the greatest lexical diversity for the three cases (number of steps in the curve), reflecting the greatest expressive richness. An indicator with a cutoff of 0.63 is suggested for HIGH.

GCN is minimal in the scope of the ID3 case, as well as that of ID5. Where the values do not reach 0.30, that would be a low rating level. This could be interpreted in terms of lexical complexity, that is, the degree of richness of expression of the information. It is interesting to see that it is true that the low levels of word repetition counts occur in these two surveys (in fact, practically the minimum of the whole set), while if it is taken from Table 2, the product of DL × SR gives the smallest values of the whole set, indicating less expressiveness of the text. It is also given that their PMD values are in low to medium values. Figure 5a effectively shows that the number of "jumps" in the frequencies is few (2 and 3).

The GCA is also one of the lowest for these same cases. It corresponds to companies where the volunteer who responds is the director; they have a defined scope at the organization level, but apparently, it is not completely achieved since they cannot answer the typical questions that describe it. They also present problems in determining the degree of scope. Both cases correspond to local companies in Concepción del Uruguay, both dedicated to software consulting.

The GCN for ID scopes 1, 4, and 6 has a value between 0.3 and 0.5. This constitutes a slightly better value than the previous one and corresponds both to texts with greater lexical diversity and the greater DL × SR product (0.0275, 0.0858, 0.077). We would be facing more elaborate, expressive, and longer texts. This would correspond to more information provided and therefore a better understanding of the business. This in turn coincides with the maximum levels of PMD in two of the three cases. The counts here are also somewhat higher for two of the three cases. Note that the ID 4 company does not have a maximum PMD but one of the highest DL.

In Fig. 5b, it is observed that the frequency jumps are greater than before (typically 3 except in case 4, which reaches its complexity in another way).

These cases all fall in the range of intermediate values of GCA; typically, they are medium and large companies located in different provinces and countries. The branches correspond to engineering companies with inputs and personnel of various profiles, typically geographically spread over more than one city or belonging to the state level. In these cases, the degree of understanding of the business is much higher than the previous case, but not complete. In contrast, the scope is more perfectly defined.

The GCN for ID scopes 2, 7, and 8 has a higher range than the previous ones (from 0.5 to 0.65), where a higher degree of elaboration and more effective expressiveness are already assumed. This is reflected in more varied texts than before (the number of jumps in the images in Fig. 5c exceeds three jumps, and the IMD is greater). It is also noted that the DL × SR is, on average, higher. It includes more mature companies in terms of internal administration. Of the cases studied, two are small and one is large. In all cases, the volunteers are with complete knowledge of projects. None of the companies have branches, but two of them are in major provincial capitals. In these cases, the degree of scoping is balanced and quite good.

As for ID 9 and 10, both present the highest values. Their maximum word count is the highest, the DL × SR value is 0.21, and the number of jumps does not fall below 5, as in Fig. 5d, showing a richness and diversity of maximum values. However, the words have a lower PMD than in other cases, thus resorting to shorter words and presumably more frequently used.

In these cases, the highest GCA level coincides, showing the highest qualification in terms of formalization. It corresponds to large companies, where the person responsible for answering the survey is the project leader. In both cases, the

Table 4 GCA indicators

ID	GCA	Indicator
1,2,3,4,5	<0.2	LOW
4,6,7,8	[0.2–0.3]	MEDIUM
2,7,8	>0.3	HIGH

Table 5 GCN indicators

ID	GCN	Indicator
3,5	<0.3	VERY LOW
1,4,6	[0.3–0.5)	LOW
2,7,8	[0.6–0.68)	MEDIUM
9,10	≥0.68	HIGH

Table 6 GCN and GCA according to the indicators

ID	GCN	CM	SR × DL	GCA
3,5	VERY LOW	≤4	<0.003	LOW
1,4,6	LOW	[3–6]	[0.03–0.08]	LOW - MEDIUM
2,7,8	MEDIUM	[5–8]	[0.01–0.2]	MEDIUM
9,10	HIGH	[8–9]	(0.2–1]	HIGH

companies do not have branches and are located in an important provincial capital. It is interesting to note that in one of the cases, the scope was not previously defined, and this is reflected in a lower GCA.

In both cases, the companies are established centers, solid in the industry and with experience in project management methodologies.

Table 4 provides a summary of the values found and their relationship with the cases. It also proposes an indicator rating for the associated value ranges.

In a similar way to what was done with the GCA values, Table 5 presents a summary of GCN and a proposal of indicators.

In Table 6, a summary of the cases is presented, along with the dominant characteristics according to GCA, GCN, and the linguistic counts of their texts. Clear delimitations are observed, which must be statistically confirmed with more cases.

An interesting fact that the study of Table 3 provides has to do with the accent that the company puts in each instance with respect to its line of business. In the case of ID 3, only "NO" appears as notable in the case of services, "Payments" appears in the case of business requirements, and "Platform" and "integration" appear in project requirements. Being a company in the software field, it would indicate that there is a product that is not totally determined, but payments for the products are intended.

In the case of ID 5, the situation is worse, as no significant term is in sight.

Regarding intermediate GCA companies, some more defined terms are observed as relevant, such as the case of ID 7, where "availability" is detected as a factor that the company has as its key characteristic.

Companies where GCA and GCN are high have much more specific word differentiations, typically aligned with their core product.

6 Conclusions and Future Work

This work presents a subset of the metrics for the evaluation of project management methodologies, focused on managing the scope of the project. Only the Business Understandability (GCN) and the Degree of Completeness of the Scope (GCA) have been used since other metrics such as Requirements Validation Criteria (CVR) and others were outside the scope of this chapter. A case study with ten companies is also presented, the metrics are applied to these cases, and certain evaluation indicators are proposed.

From the preliminary results presented here, it can be stated that the metrics allow the identification of descriptions of scope and requirements at different levels of specificity, which are compatible with simple and traditional linguistic analyses.

Among the pending tasks are the formulation of metrics for the management of the other areas of knowledge that must be considered in project management, such as time management, cost management, quality management, resource management, stakeholder management, communications management, risk management, and procurement.

It also remains to check the linguistic relationships between and metrics with the largest number of cases. It also remains to evaluate not only the characteristics of the management documents but also the impact of quality in the present and future exercise of projects.

Another interesting point is to relate the metrics for the identification of the implicit knowledge found within the texts that describe the scope of the project, applying text mining techniques, referring to the process of discovering and extracting relevant and nontrivial knowledge from unstructured texts. This is relevant given that the complexity of natural language makes it difficult to access information in texts and it is still far from being able to construct general-purpose representations of meaning from text without restrictions.

References

1. Project Management Institute (2017) A guide to the project management body of knowledge. 6th edn. ISBN- 10: 9781628251845
2. PRINCE2 (2009) An introduction to PRINCE2: managing and directing successful projects. Office of Government Commerce. Stationery Office, 123 p. ISBN-10: 0113311885, ISBN-13: 978-0113311880
3. Böhm A (2009) Application of PRINCE2 and the impact on project management. ISBN (eBook) 978-3-640-42634-8

4. Highsmith J (2010) Agile project management: creating innovative products, 2nd edn. Addison-Wesley, Boston. 432 p
5. ISO (2012) ISO 21500:2012 guidance on project management. ISO, Geneva
6. Sutherland J (2014) Scrum: the art of doing twice the work in half the time. Crown Business, New York. 256 p. ISBN-10: 038534645X, ISBN-13: 978-0385346450
7. Van Solingen R, Van Lanen R (2014) (Scrum for managers) Scrum voor Managers. Academic Service, Den Haag. EAN: 9789012585903
8. Lei H, Ganjeizadeh F, Jayachandran P, Ozcan P (2015) A statistical analysis of the effects of Scrum and Kanban on software development projects. Robot Comput Integr Manuf. https://doi.org/10.1016/j.rcim.2015.12.001
9. Shearer C (2000) The CRISP-DM model: the new blueprint for data mining. J Data Warehousing 5(4):13–22
10. Shafique U, Qaiser H (2014) A comparative study of process models data mining (KDD, CRISP-DM and SEMMA). Int J InnovSci Res 12:217–222
11. Varajão J, Dominguez C, Ribeiro P, Paiva A (2014) Critical success aspects in Project management: similarities and differences between the construction and software industry. Tech Gazette 21(3):583–589
12. TSG (2018) The CHAOS report. The Standish Group. Disponible en https://secure.standishgroup.com/reports/flyers/CM2018- TOC.pdf
13. Lehtinen T, Mäntylä M, Vanhanen J, Itkonen J, Lassenius C (2014) Perceived causes of software project failures – An analysis of their relationships. Inf Softw Technol 56:623–643
14. Ramos P, Mota C (2014) Perceptions of success and failure factors in information technology projects: a study from Brazilian companies. Procedia Soc Behav Sci 119:349–357
15. Montequin S, Fernandez C, Fernandez O, Balsera J (2016) Analysis of the success factors and failure causes in projects: comparison of the Spanish Information y Communication Technology (ICT) sector. J Inf Technol Project Manage 7:18–31
16. Chow T, Chao D (2008) A survey of critical success factors in agile software projects. J Syst Softw. Available: Science Direct 81:961–971
17. Elkadi H (2013) Success and failure factors for e-government projects: a case from Egypt. Egypt Informatics J 14:165–173
18. ElEmam K, Koru A (2008) A replicated survey of IT software project failures. IEEE Software 25:84–90
19. Blaskovics B (2016) The impact of project manager on project success – The case of ICT sector. Assistant Professor, Strategy and Project Management Department, Corvinus University Budapest E-mail: balint.blaskovics@uni-corvinus.hu
20. Esteki M, Gandomani T, Farsani H (2020) A risk management framework for distributed scrum using PRINCE2 methodology. Bull Electr Eng Inf. https://doi.org/10.11591/eei.v9i3.1905
21. Jabar M, Mohd AN, Jusoh Y, Abdullah S, Mohanarajah S (2019) A pilot examination of an improved agile hybrid model in managing software projects success. Test Eng Manag 28:3040–3046
22. Smoczyńska A, Pawlak M, Poniszewska-Marańda A (2019) Hybrid agile method for management of software creation. In: Advances in intelligent systems and computing. Springer, Cham (Denmark), pp 101–115
23. Wysockia W, Orłowski C (2019) A multi-agent model for planning hybrid software processes. Procedia Comput Sci 159:1688–1697
24. Mousaei M, Gandomani T (2018) A new project risk management model based on Scrum framework and Prince2 methodology. Int J 9:442–449
25. Cuccurullo C, Aria M, Sarto F (2016) Foundations and trends in performance management. A twenty-five years bibliometric analysis. Scientometrics 108:595–611
26. Van Eck N, Waltman L (2010) Software survey: VOSviewer, a computer program for bibliometric mapping. Scientometrics 84:523–538

27. Barclay C, Osei-Bryson K (2010) Project performance development framework: An approach for developing performance criteria & measures for information systems (IS) projects. Int J Prod Econ 124:272–292
28. Ivan I, Ciurea C, Zamfiroiu A (2014) Metrics of collaborative business systems in the knowledge based economy. Procedia Comput Sci 31:379–388
29. Kumar L, Misra S, Rath S (2017) An empirical analysis of the effectiveness of software metrics and fault prediction model for identifying faulty clases. Comput Stand Interfaces 53:1–32
30. Arar Ö, Ayan K (2016) Deriving thresholds of software metrics to predict faults on open source software: replicated case studies. Expert Systems with Applications 61:106–121
31. Wanderley M, Menezes J, Gusmão C, Lima F (2015) Proposal of risk management metrics for multiple project software development. Procedia Comput Sci 64:1001–1009
32. Vandevoorde S, Vanhoucke M (2007) A simulation and evaluation of earned value metrics to forecast the project duration. J Oper Res Soc 58:1361–1374
33. Vanhoucke M (2011) On the dynamic use of project performance and schedule risk information during project tracking. Omega 39:416–426
34. Khamooshi H, Golafshani H (2014) EDM: Earned Duration Management, a new approach to schedule performance management and measurement. Int J Proj Manag 32:1019–1041
35. Wood D (2017) High-level integrated deterministic, stochastic and fuzzy cost-duration analysis aids project planning and monitoring, focusing on uncertainties and earned value metrics. J Nat Gas Sci Eng 37:303–326
36. Basso D (2014) Propuesta de Métricas para Proyectos de Explotación de Información. Revista Latinoamericana de Ingeniería de Software:157–218, ISSN 2314-2642
37. Papa M (2014) Aseguramiento de la Calidad de un Recurso Organizacional: Evaluando y Mejorando una Estrategia Integrada de Medición y Evaluación. Tesis Doctoral. Facultad de Informática, Universidad Nacional de La Plata
38. Becker P (2014) Visión de proceso para estrategias integradas de medición y evaluación de la calidad. Tesis Doctoral. Facultad de Informática, Universidad Nacional de La Plata, pp 1–2012
39. Rivera M (2018) Enfoque Integrado de Medición, Evaluación y Mejora de Calidad con soporte a Metas de Negocio y de Necesidad de Información: Aplicación de Estrategias a partir de Patrones de Estrategia. Tesis Doctoral. Facultad de Informática, Universidad Nacional de La Plata
40. Arvanitou E, Ampatzoglou A, Chatzigeorgiou A, Galster M, Avgeriou P (2017) A mapping study on design-time quality attributes and metrics. J Syst Softw 127:52–77
41. Chillarón M, Quintana-Ortí G, Vidal V, Verdú G (2020) Computed tomography medical image reconstruction on affordable equipment by using Out-Of-Core techniques. Comput Methods Prog Biomed 193:105488
42. Tebes G, Peppino D, Becker P, Matturro G, Solari M (2020) Olsina analyzing and documenting the systematic review results of software testing ontologies. Inf Softw Technol 123:106298
43. Jaleel A, Arshad S, Shoaib M, Awais M (2019) Design quality metrics to determine the suitability and cost-effect of self-* capabilities for autonomic computing systems. https://doi.org/10.1109/ACCESS.2019.2944119
44. Friedman A, Flaounas I (2018) The right metric for the right stakeholder: a case study of improving product usability. https://doi.org/10.1145/3292147.3292224
45. Raza B, Aslam A, Sher A, Malik A, Faheem M (2020) Autonomic performance prediction framework for data warehouse queries using lazy learning approach. Appl Soft Comput 91:106216
46. Yazar A, Arslan H (2018) A flexibility metric and optimization methods for mixed numerologies in 5G and beyond. https://doi.org/10.1109/ACCESS.2018.2795752
47. Bracy KB, Seddon E, Tommbs T (2019) A framework for evaluating biodiversity mitigation metrics. https://doi.org/10.1007/s13280-019-01266
48. Paschali E, Ampatzoglou A, Escourrou R, Chatzigeorgiou A (2020) A metric suite for evaluating interactive scenarios in video games: an empirical validation. In: Proceedings of the ACM symposium on applied computing 30

49. Benmakrelouf S, St-Onge C, Kara N, Tout H, Edstrom C, Lemieux Y (2020) Abnormal behavior detection using resource level to service level metrics mapping in virtualized systems. Futur Gener Comput Syst 102:680–700

50. Le N, Hoang D (2017) Capability maturity model and metrics framework for cyber cloud security. https://doi.org/10.12694/scpe.v18i4.1329

51. Kbaiera W, Ghannouchib S (2019) Determining the threshold values of quality metrics in BPMN process models using data mining techniques. Procedia Comput Sci 164:113–119

52. da Costa J, de Souza A, Rosário D, Cerqueira E, Villas L (2019) Efficient data dissemination protocol based on complex networks' metrics or urban vehicular networks. https://doi.org/10.1186/s13174-019-0114-y

53. Ahmad M, Odeh M, Green S (2018) Metrics for assessing the basic alignment between business process and enterprise information architectures with reference to the BPAOntoEIA framework. IEEE 978-1-7281-0385

54. Beecks C, Grass A, Devasya S (2018) Metric indexing for efficient data access in the internet of things. In: IEEE International conference on big data 978-1-5386-5035

55. Kapur R, Sodhi B (2020) A defect estimator for source code: linking defect reports with programming constructs usage metrics. https://doi.org/10.1145/3384517

56. He P, Li B, Liu X, Chen J, Ma Y (2015) An empirical study on software defect prediction with a simplified metric set. Inf Softw Technol:170–17s

57. Gopal M, Amirthavalli M (2019) Applying machine learning techniques to predict the maintainability of open source software. https://doi.org/10.35940/ijeat.E1045.0785S319

58. Zagane M, Abdi M, Alenezi M (2020) Deep learning for software vulnerabilities detection using code metrics. https://doi.org/10.1109/ACCESS.2020.2988557

59. Ismail S, Mohd F, Jalil M, Wan Kadir W (2019) Development metrics measurement level for component reusability evaluation approach (CREA). https://doi.org/10.11591/ijece.v9i6.pp5428-5435

60. Oliveira B, Da S, Martins C, Magalhães F, Góes L (2019) Difference based metrics for deep reinforcement learning algorithms. https://doi.org/10.1109/ACCESS.2019.2945879

Blockchain Technology Applications for Next Generation

Neha Puri ⓘ, Vikas Garg ⓘ, and Rashmi Agrawal ⓘ

1 Introduction

The term assigned to the infrastructure promoting the digital currency of Bitcoin is "blockchain technology." In a certain structure, contemporary digital currencies are centralized in order to give liquidity of electronic data and are only exchanged in an area under the authority of the mechanism. Conversely, Bitcoin data are handled in a shared system. This involves the use of a system in which the value of electronic data is widely understood by all people while preserving its value. This revolutionary framework is compatible with blockchain technologies. The event in which Mt. Gox, a Bitcoin exchange, lost customer-owned coins in February 2014 did not compromise the reliability of Bitcoin itself or its technologies. This is no different from the situation where the money is stolen from an exchange firm and the company goes bankrupt. The value of the currency would not be impaired.

This chapter analyzes the implementation of blockchain technology, information exchange, and other facets of the Bitcoin system, the hurdles that are yet to be solved, and our efforts to solve these obstacles. It also tackles our participation in the distributed open-source Hyperledger project.

N. Puri · V. Garg (✉)
Amity University, Noida, Uttar Pradesh, India
e-mail: npuri@amity.edu; vgarg@gn.amity.edu

R. Agrawal
MRIIRS, Faridabad, Haryana, India

© Springer Nature Switzerland AG 2022
P. Raj et al. (eds.), *Blockchain, Artificial Intelligence, and the Internet of Things*,
EAI/Springer Innovations in Communication and Computing,
https://doi.org/10.1007/978-3-030-77637-4_4

2 Summary of Blockchain Innovations

The method by which the blockchain works in the Bitcoin ecosystem is initially discussed in this chapter. In Bitcoin's context, a concept analogous to a bank account in a standard banking transaction is considered an "address." The input and output address and the amount of Bitcoin are part of a Bitcoin transaction [1]. A data gathering of a sequence of legal transactions recorded every 10 minutes is referred to as a block, and a blockchain refers to a number of chronologically connected blocks. New blocks are created and managed by "miners." A miner who produces a block below a certain threshold with a hash value earns Bitcoin as a reward. Here is the "race" of the miners to make bricks. The hash value threshold is set in such a way that, on average, a race takes place every 10 minutes. If one of the following blocks is recalculated, the hash value is substituted into the following row so that it is not possible to guarantee consistency when the old data are modified. Therefore, it's almost impossible to tamper with a blockchain. Besides, all the nodes that join the nodes that join the P2P Bitcoin network share a unique blockchain, thereby simultaneously realizing high availability. Although Bitcoin transactions are independently handled, they are merged into a distinctive and consistent blockchain to build a decentralized tamper-resistant and highly accessible information system (ledger) [2].

3 Application of Blockchain Technologies to Sharing Knowledge

Bitcoin's blockchain deals with Bitcoin transactions as information only. In theory, however, it can handle a lot of other types of data. Second, a case study is carried out in which the blockchain as a knowledge-sharing platform is applied to the business phase. A joint verification trial with Mizuho Bank on cross-border movement of securities was done. Owing to the complex procedure involved, the average cross-border exchange in shares takes 3 days from execution to payment. This is because there is a need for a long time to verify settlement orders and implementation content [3]. In order to avoid risks such as market instability, it is very important to restrict this time. Previous experiments have looked into time saving by the use of unified administration data sharing, but this has not been done due to high system operation and management costs. The joint verification experiment with Mizuho Bank set up a system that documented an execution case in a block using the Open Assets Protocol. Both parties interested in the agreement have agreed that information on the execution (which has been tamper-resistant information) will be shared within a short amount of time. It emerged that the entire procedure could be reduced to less than 1 day instead of 3 days.

3.1 Key Categories

We take a look at how blockchain is spreading to these markets.

4 Finance

4.1 With Confidence Crowdfunding

There's an accountability issue for crowdfunding today. Why? Why? As many as 85% delayed distribution, while 14% failed to deliver what had been planned. So what happens if you get funding from the wrong campaign and someone misuses the money? You will never want to fund a technology initiative again. Through blockchain software, you will understand further information, like who you're going to give the money and how the money is going to be used by the developers. On the other hand, with lower rates and overall costs, without excessive fees on the part of lawyers, developers may secure more support for their projects. It's almost like renting a house, except you don't give up all the money in advance. In Escrow, you're holding a currency. Only after the estate developer builds a house for you will a portion of the funds be released. Similarly, with the assistance of a smart contract, crowdfunding funds will not be released until the developer has made progress on the project. Yet, you have the confidence [4].

Backer Insurance

Backers	Milestone 1	Milestone 2	Milestone 3	Creators
	Vote	Vote	Vote	
Refund			Rewards	

Pledgecamp is the Latest Wave Crowdfunding Destination. Unlike the new versions launched by business leaders Kickstarter and Indiegogo, this platform is a stable and accessible blockchain-powered crowdfunding framework.

4.2 Cash Move Abroad

When you need to pass smaller sums abroad, what do you do? Maybe, you're using a transfer program like Western Union. This trustworthy middleman moves money worldwide between its offices. Finally, using overseas money transactions, it balances the accounts on the backend. These programs, though, also charge you a premium of 10% or more [5, 6]. But now, blockchain financial systems are changing the way money is exchanged across the globe at a lower cost. Bitcoin is the first-use example of this kind. Companies like Ripple, built on this radical financial technology, are changing the way money is distributed to everyday people around the world.

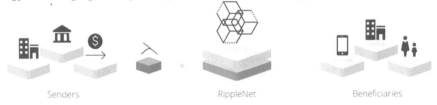

Senders   RippleNet   Beneficiaries

4.3 Enable Fast Loans and Lend Money

Have you ever tried to lend your money to a bank? Let's face it: the current lending market is unsustainable, especially with a lot of paperwork to fill in and a high interest rate. P2P lending is spreading rapidly in personal finance. Today, via blockchain, both borrowers and lenders are connected around the world, without giving a break to the bank. You will remain absolutely anonymous as lenders and do not need to report anywhere. Only choose your preferred loan from the marketplace. First of all, get your wallets started. In a low-interest setting, creditors are going to repay the loan. Decentralized by 100%, Lendoit is dismantling the mainstream lending industry. Loans from the industry are projected to cross $290 billion by 2020.

4.4 Effective Foreign Exchange Postprocessing

The world's biggest and most liquid asset groups are foreign exchanges. The capture, mediation, and resolution of trade operations are difficult. Axoni's network offers completely synchronized collaboration with industry data suppliers from third parties, thus improving performance in the overall industry.

4.5 Stock Market in Blockchain

Imagine businesses can log, question, and exchange on a blockchain-based network. What does this mean to you? In a nutshell, you can buy business stock in digital currencies. London Stock Exchange is checking this by selling wireless private shares of small and medium-sized businesses in Italy. In addition, Overstock, an online retailer, began issuing company shares through Bitcoin blockchain technology.

4.6   Increasing Central Bank High Uptime

The use of a cashless service is popular, for example, payment by credit card, Apple Pay, or QR code. Still, what happens when the computer goes down for a long time? The fact is that Bitcoin has been going on for a long time and has not yet suffered a nanosecond of downtime. As a result, a rising number of central banks are actively analyzing blocks.

4.7   Reduce Fraud in Insurance

Numerous fraud systems may be revealed to insurers. For instance, by withholding vital information, a new applicant may commit fraud, or, on behalf of disqualified dependents, by making a lawsuit. But what should we do to reduce insurance abuse? The blockchain network tracks medical treatments and time stamps. This facilitates the compliance and authentication of medical facilities provided in accordance with IBM and Cantina Technology; 15 Indian insurers are bound together to build this blockchain-based approach.

4.8   Digital Wills for Transition

The number of people who use cryptocurrencies to store funds is continuously rising. So, if you die without a wish, what happens to your cryptocurrency? How do you avoid the depletion of your savings and encourage your loved ones to move them? By using My wish, during a crisis, you build smart contracts to manage your funds. This helps you to move the assets of your family or friends in the event of a severe illness or death.

4.9   Capital Price Forecast

One of the main problems with emerging forecasting markets is that they are concentrated. As a consequence, unified platforms have a single form of failure. So they're quicker to shut off. This requires a credible person to report in such a formal template honestly and correctly each time. But what happens if this individual makes mistakes, is immoral, or manipulates the results? [7, 8].

5   Power

5.1   Lower Home Electricity Bills

A smart plug (proof of concept) was introduced by Accenture to manage power consumption and save costs. First, the prototype communicates with other gadgets using power monitoring. Next, the lowest energy tariffs are being checked. Finally, blockchain is used to switch vendors quickly. As a result, this helps many consumers who pay a meter at reduced pay for their electricity.

5.2   Solar Electricity Trading with Neighbors

Are you able to use your power at peak hours, when the price is the greatest, to charge a battery? Or, do you want to use solar power at a lower cost to do the same task? This project, run by LO3 Energy, aims to enable people to *buy and sell renewable energy* to their neighbors.

That's how it works. While solar panels are soaking up the sun on the roofs of terraced houses, the amount of computers attached to the panels is still crunching. Next, they count the number of electrons released. Then, they're going to write the number on a blockchain. In this way, you can trade in solar power and circumvent the sale of electricity to the utility company [9].

6   Healthcare

6.1   Secure Medical Records

Today, the standard of treatment they can offer is restricted by physicians, nurses, and health practitioners. You will also post the medical history anonymously on the cloud for blockchain apps in healthcare. And, be confident that it can be used anywhere in the world by you or an authorized official. Estonia's government is working with one of the hardware firms in the blockchain, Guardtime, specifically to create a chain of citizen health information [10]. This allows people to hold their ID credentials to unlock access to their health records in real time.

6.2   Check for Fake Medicine

If the medication is real, you wouldn't know until you inserted it into your stomach. Blockverify offers a solution in the supply chain to monitor pharmaceuticals and to ensure that an authentic product is purchased by customers.

7   Real Estate

7.1   Registry of Land Ownership

Maintaining a land ownership nation register is not only costly but also a labor-intensive process. The first blockchain safe application for record-keeping of real estate is Ubiquity. It records and tracks the ownership and other records of the house. Also, it decreases potential quest time for titles and increases accountability. Government partners of Sweden with ChromaWay Here are planning to test the probability of a blockchain-based land registry.

8   Tourism

8.1   Give Tourist Coupons and Promotions

If you like it or not, companies provide their customers with traditional paper, pre-paid passes, coupons, discounts, and other types of inefficient benefits and loyalty rewards. Rouge helps corporations to make use of this modern digital format to create their Blockchain loyalty tokens and coins for their clients through their customer service systems.

8.2   Fill Up Empty Hotel Rooms

If you're a sales agent in the hotel business, you'll love this approach. Imagine the hotel guest canceling his long stay at the last minute and he declines to pay to make the matter worse. In comparison, locating a replacement client for space within a limited period is very difficult. Webjet is an online travel portal that enables empty hotel rooms to be stored.

9 Social

9.1 Collect Friends' Points

You are young now and you have a deep passion for building a new startup to change the world. That's brilliant. Yet, there is one concern you have. You don't have immense money. How do you persuade the venture partners that you are resourceful and trustworthy? Before they fund your startup, are you a helpful person? A way for you to earn points from your mates is the Masachain blockchain game. They grant points to you anytime your mates perceive a good feeling from you. Support, appreciation, and kindness are alluded to in these optimistic emotions. With this platform, your venture capitalist or an outsider can trust you better. And there's no need for a second-hand man or to dig into your social media profile to find out what kind of person you're out of work [11, 12].

9.2 Manage User Reputations

Recently, Airbnb started tests using a blockchain as a method of managing the trust of customers. The business reacts to feedback provided by both tenants and landlords. As a consequence, how trustworthy future hosts or visitors can be will be judged by the client.

10 Marketplace

10.1 Advertising Inventory for Buy and Sale

Are you tired of paying high ads prices online? Or how about being afraid of the ban on online content? Thrive is a rewarding blockchain site where advertisements are bought at extremely low rates by buyers and sellers. You will also be paid if you want to search the websites and share details.

10.2   Know All About the Things You're Purchasing

Look beyond the label

You just bought a high-quality wallet, and you're wondering if it has more detail. What if you could peer beyond the mark to reveal the factual facts of the work, the origins, and the influence of the creators? Provenance is provided with a special ID for physical items. Through means of an ID, you gain access to a safe digital history of how the product is distributed to your business.

10.3   Get Rid of Commercials by Banner

Isn't it disturbing to see a lot of web banners as you read the chapter? And you're without them, huh? Strong opinion allows advertisers to show ads to audiences that are interested and targeted. This is how it works: the advertiser positions similar ads right above the title of a great story.

10.4   Zero Costs for the Platform

Sellers typically have to pay certain fees on the website for the common online marketplace. These payments are, of course, either absorbed by vendors or handed on to the customers. The open bazaar is a special online platform that recognizes the only cryptocurrency of its type. As there are no intermediaries, it is a peer-to-peer application without costs and limitations. You're in charge of your shop, and you're in control.

11 Identity

11.1 Tell the World About the Identity You Possess

The way we prove our identities to others remains based on paper and plastic documents, physical contact, and handwritten signatures. Today, as we engage online and offline, our identity data are retained by individual organizations. Data are also poorly integrated and out-of-date. This duplication is vulnerable and inefficient to error and poses safety risks. There should be a safer way to secure our identities with master identifying papers. This feature is controlled directly by each user. With a digital identity network, Deloitte blockchain apps provide a solution that supports both privacy and protection.

12 Mathematics

12.1 Solving Difficult Questions in Mathematics

The Riemann hypothesis is a complex question of mathematics that has yet to be proven. This remains one of the field's excellent puzzles. This math problem includes the discovery and evidence of major supercomputers. Riecoin, another cryptocurrency, set out to address the issue of prime numbers. A lot of mathematicians predict the Riemann hypothesis to be ultimately solved with the assistance of Riecoin.

13 Transport

13.1 Location Data

They can verify the shipping position details when you order a product online. High-level data on key locations such as airport transport and collection centers are expected to be included in these specifics. You can't know the specifics of the exact location [13, 14]. And if the centralized system crashes, what's going to happen? XYO Network introduces location data from several interconnected devices to address this issue, which will provide the details required to execute smart contracts. Use cases are very useful for companies who need place info. Airlines, for example, need to locate missing bags; car rental agencies need to trace misplaced keys or e-commerce [15].

13.2 Parking Space Rent Out

In a busy place, imagine you need a parking spot. By anyhow you could find a place to park with advance payment done for that, wouldn't that be good? Parking owners may use blockchain to place their parking lot on rent when they are out of time or on vacation, which in turn would be easy for the one who is search of the parking space. Airbnb parkgene aims to be made available for a temporary parking facility.

13.3 Support Car Drivers to Pay Parking Fines

Thinking of a driverless automobile is a fact now, but on this condition, it doesn't mean that a vehicle deserves free parking without a driver. The parking fines still have to be charged by someone. The conventional bank account is not held by these vehicles. But, in the blockchain, they will get one. With Car Wallet, when you submit money to her wallet, the car is told. Then all you need to do is opening the door, authorizing the beginning of the engine, and paying the parking fees.

13.4 Hire a Ride

Users have to submit their driver's license and fill out some paperwork when they rent a car. The moment you submitted the money and there's you get the key to the car. To you, that's maybe convenient, isn't it? But if you look at the whole process of car rental, it is slightly complicated. It may be or it is just a car that we think that car makers make leasing seem too quite easy. The challenges faced by automobile leasing firms are that there is no upgrading of details. Imagine just using a blockchain. The latest information can be tracked, exchanged, analyzed, and updated by any individual in the car rental supply chain. Best of all, regardless of where the car is in the lifecycle, you get the new results. With IBM blockchain, less processing time, reliable details, and decreased overhead costs are enjoyed by the entire car rental supply.

13.5 Paddle Riding in Real Time

Nowadays, it is popular to hitch a lift. And, with either money or credit, the passenger pays the driver. What about the currency of digital? LaZooz is a real-time ridesharing app operated by the group, powered by blockchain. In real time, this network synchronizes empty seats with passengers and matches like-minded individuals. So, what's the good news? Both drivers and riders are rewarded with digital incentives.

13.6 Car Sharing from Peer to Peer

What happens if Uber tries to move out of the city? Many drivers are out of job, a major hit on tourism and travel. Using blockchain technology for ride sharing and hiring can be a new effective way. Arcade City strives to beat Uber. They win Arcade Tokens until all riders and drivers are directly linked, and they can either sell for cash immediately or use for servicing.

13.7 Going to Space

Have you ever dreamed about going to a galaxy or a space on a vacation tour? Virgin Galactic is currently not sending people to orbit, but tickets are being sold. It costs $250,000 to enter a queue, and you can pay the amount in Bitcoins.

13.8 Charge Lower for Electric Car Charging

Your account for the total amount of time vanished connecting to the car to the charging station to drive an electric car. Why? Why? And the precise power consumed cannot be measured under the old billing scheme. And, this amount of time we've spoken about normally takes hours. The goal is to connect the charging stations of electric vehicles to blockchain technology. The docking station gives user *authentication, payment processing, and loyalty point* assignment. This will actually lead to simplifying the complete process of consumption of energy in the energy market arena. There are many automobile charging stations across Germany that have been opened by RWE, a German energy firm. These are connected to the shared blockchain of Etheraum. Today, in micro-transactional slices, consumers can charge their vehicles and pay.

13.9 Reducing Paperwork for the Shipping Industry

In container shipping, margins are still small. Yet, documentation handling is more expensive than shipping containers. Maersk, the leading provider of container ships, and IBM have created a joint partnership leveraging blockchain technology and its advancements. They are aimed at making global trading more efficient, more open, and more stable.

14 Entertainment

14.1 Watch a Single Football Match: Pay Per Use

The monthly membership rate model for flats is loved by businesses. About why? There is a recurring and stable gain. So why can't we pay to watch a single soccer match instead of a whole soccer channel? The channel sources and providers provide viewers with an opportunity to settle on the pay-per-use basis for multimedia services using Raiden. This payment is made through Ethereum micropayments.

15 Rights Protection

15.1 Security of Logo for Wine

If you're a wine fan, this is what you'll enjoy. Do you know the researchers say that 20% of the wine drunk globally is counterfeit? They refill these phony wines and sell them to us. Not only have these forgeries caused an estimated decline of nearly about 2 billion in sales, but also, most specifically, consumers have now much less interest in high-class branded wine. The cryptography framework was developed by catalogs to protect the identification of bottle tops and label on the wine. The validity can be verified with the given seal by the consumers via the smartphone easily.

15.2 Preserve Fine Arts Developed by Talented Artists

It is not impossible to clone fine arts from a museum. What you need is to take a snapshot to get it to be replicated by an online freelancer. Increasingly, consumers are suspicious of art forgery.

With encrypted microchips inserted in the paintings, chronicled created a solution and recorded with the blockchain. As a consequence, if the artwork is authentic, buyers will say.

15.3 Protect the Luxury Footwear

Do you know that the most frequently falsified thing is footwear? Have you ever asked how a bogus maker gains a much cheaper price by selling sneakers? Well, when they bypass production expenditures such as initial design and testing, marketing, and legislation, these factories gain massive profits. It was hard for the sneaker experts to differentiate fakes from actual ones. You should take out your mobile and check the microchip in the sneakers with Chronicled's solution. And, you'll know whether this (Michael Jordan's) sneaker is genuine or false.

15.4 Offer a Digital Identity to Your Valuables

It is a traumatic feeling to lose your valuables. It's not sufficient to give the investigators the appropriate identification number of your lost valuables. You will preserve what you enjoy at VTT whether it's a bike, a drone, a scooter, or a luxurious handbag. To the Ethereum blockchain, you write the proofs of ownership forever. In this way, the retrieval of these compact stolen goods makes it faster.

16 Government

16.1 Smart City

If the globe adopts 5G technologies, you might assume that the networks of a country will remain more linked to "intelligent" smartphones. The lampstands, security cameras, retail stores, kiosks, or buildings are these 'smart' gadgets. They're referring to one another. What does this mean to you, then? For 1 kg of prawn, assume you might know which grocery and supermarket will provide you with the lowest price. As you have real-time traffic alerts including collisions, you also have the shortest road and the available spaces for parking. If you take public transit, the time of the next bus is exact. This detail makes it more effective and fun for your journey. These effective things are possible due to sensors installed in the physical environment.

16.2 Give Untampered Votes in Elections

Any traditional methods of voting in a country are via a ballet-paper method or electronic voting via a polling station installed in a computer or i-voting through the web. Confusion regarding the capacity of the polling mechanism to ensure cybersecurity and data privacy and protect against future threats is still present. Each count of votes, after all, has a big effect on a government. To ensure fair and verifiable polls across the globe, Voatz created a voting forum. Voatz offers a test smartphone blockchain voting for mobilized military forces and overseas US residents in preparation for the 2019 Local Elections. The effective return rate for the ballot obtained was 98%. There was a high 90% mobile voter turnout figure. We're not far from using this technology,

17 Charity

17.1 *Donate Funds to Help the Unfortunate*

You want to donate to a good cause with funds. And, you are unclear if your contribution would go directly to the poor. For charitable fundraising, Bitgive ensures accountability. In real time, you get to see the donation results. Dogecoin was also praised as a good coin to collect charitable funds. This assertion was put to the test by Dogecoin backers in 2014 when they raised $30,000 to help send Jamaican to Sochi Olympics.

18 Automation

18.1 *Remotely Automated Robots*

AI is known for its numerous advantages including learning, logarithmically and intelligently designed for handling a thoughtful series of activities. You should outsource these tasks when tasks are too difficult for AI. Freelancers can monitor the activities of robots as "pilots" from a distance. This enables remote staff to produce additional cash with the Aitheon platform by controlling robots to execute complicated tasks and improve automation.

18.2 *Host Your Site Without Any Central Server*

When your server is down, you need website support. The outage is due to server congestion in most instances. Would it be nice if there were no central servers needed for your online content? Zero net creates a fully decentralized server and online connectivity. The main advantage is that it helps banish the need to host websites on servers by seta siding website data secretly in tiny parts around the Internet.

18.3 *Your Hard Disk Space Rental*

In the long term, paying for cloud computing is costly. Although uptime is reasonably good, due to outages and DDoS assaults, there is a risk of downtime. They actually can't do much, whether you choose to invest in some external services by IT or call an expert. These things are not entitled to full admin powers. Storj.io is a community sharing cloud that promises reduced costs and a higher degree of security and stability. The advantage of the Store platform is its extra special ability to make use of a storage node's free space, be it a single operating PC or large data centers. The best part is that you can still monetary some amount by renting out hard disk space for your free unused system.

18.4 Tools to Build Apps for Blockchain

Because of inadequate blockchain technology, the industry is sluggish to embrace blockchain.

To generate further software, Settlement and Cypherium set to provide with the best opportunity to act as a fundamental development unit for developers.

18.5 Get the Ink Cartridge Printer to Order

In such a scenario, you start tearing your hair out: it's late at night. When you are about to print a very significant text, the printer runs out of ink. The computers IBM and Samsung imagine will speak to each other. Just imagine it. Your printer uses smart contractual system to place a request for ink cartilage supplier for fresh stock before it vanishes off. All the integrated processes become easier and effective and open multiple ways to new business venture and partnerships.

19 Human Resource

19.1 Recruit New Fresh Grades with Trust

Fake news is there. Yes, there are counterfeit certifications. You know that if you do recruit a fresh grad, the whole recruiting process will be sluggish and painful. How can you check the credential of the recruit is genuine? One by one, get in touch with the educational institution. There should a free database and an protocol where certifications are arranged and written safely. Imagine a world in which certifications were safely written, and the chances of getting removed and altered get resolved. This is not only convenient for the organization but also for the employees as well for the free, fair, and transparent recruitment process. The smart chain provides this interesting concept of getting digital certificates for education.

19.2 Hire Staff for the Experience

Will his resume be fabricated by your prospective new hire? Again, think about it. In the resume and LinkedIn, previous employment activities are all written by the career seekers themselves. Until the time the HR himself puts the effort for contacting any former boss, you wouldn't know if it's true. For employers, here's the positive stuff. Echolink seeks to supply the work market with genuine clarity. This startup provides with a solution of genuine and verified candidate profiles with relevant skill and experience, which in turn saves time for recruiters.

19.3 Get Quick Assistance from Nearby Certified Employees

The overall structure of the economy is based on demand. You need specialist support urgently when the air-con is leaking. A large proportion of the service sector is owned by firms like Uber, Upwork, and Airbnb, and thus no further assistance is required. How amazing it would be if you could get services from a trained expert nearby that could easily get the job done? Connect Job focuses on on-demand services and creating a better way and proper mode for recognition.

20 Digital Content

20.1 Image Ownership Protection

You click a terrific snapshot of the dawning sun. The image will be widely used without your permission the moment you put it on your blog. In reality, the issue of copyright infringement and copying is faced by photographers all over the world. Well, the regulation and rights of photographs have become problematic, and sometimes it goes for the stock photography (e.g., Shutterstock, istock). On this website, photographers record their work first. Next, without their consent, they get surveillance to ensure that their images are not used publicly. Finally, immediately, photographers get license fees.

20.2 Pay Musicians on Time

The central arrangement between artist arrangements, fees, licenses, and copyrights remains the same across the entire music movement. And so, who is benefiting the most? It supports record labels and publishers of artists. The paying out of royalties to musicians and composers takes months and years. Revelator lets musicians create a quicker payment and royalty exchange.

20.3 Pay Digital Content Creators

Musicians, filmmakers, and authors are the producers of multimedia art. They create amazing content. Yet, not all are well paying. Smoogs provide content writers with a simple means of getting paid for what they do. As for the viewer, for what they buy, they just pay.

20.4 Ease Payments for Music Companies

The goal is to allow online artists and distributors to exchange information about copyright and royalties for music tracks. Any time they are played, the smart contract automatically receives recording and accepts payment as well. Ujomusic, a startup, is seeking to use the Ethereum blockchain transforming system to transform remittance for the entire music industry.

21 Retails, F&B

21.1 Loyalty Incentives for Trading

Quite frequently, with loyalty points from numerous firms, you are compensated. And, what's the drawback? For another business, you can't use these hard-earned points. Loyal strives to encourage customers to merge loyalty benefits and exchange them. More advanced loyalty programs can be sold by merchants.

21.2 Give a Gift Card to Your Friend

Your buddy's birthday is coming. You could get her a digital currency e-gift card. Block point helps merchants to create payment mechanisms for gift cards originating from the blockchain.

21.3   Discover the Lamb Meat's Origin

There has been an increase in false alarms in the food industry, leaving consumers frustrated, and then its easy flow gets into trouble. Choosing the path of freshly produced with provenance in appropriate time. If you are the one who is conscious of the root, you will tell if the lamb chop is contaminated.

21.4   Buying a Pie

Laszlo Hanyecz persuaded someone in April 2010 to send a few pizzas at 10,000 Bitcoins; that's big. The same pizzas at the end of 2016 would be worth $3.5 million when Bitcoin was selling over $700 each.

22   Business

22.1   Accurately Bill for Consultation

It's difficult to determine the payments by time if you have consultancy by phone calls. If she consults you for just 10 minutes, then the normal going 1 hour consulting charge would be okay to your client? Experty offers a way to connect two or more individuals with an online voice or video call to share paying information. Rather than taking per block as per hour, as per the prices fixed by the consultant for each minute. Depending upon the satisfaction level you got during the consultation and if both parties are convinced, then refund can be made. Otherwise, payment is deducted via the Ethereum blockchain at the moment.

22.2   Protect Your Patents

This will cost up to $3000 when you do a specialist search. Have I told you the expense of a single search? And once you intend to apply, it takes an average of 16 months for the whole patent process to be submitted for release.

23 Education

23.1 Create Wise Legal Contracts

Let us say you ought to buy an industrial office. Money and deeds, such as a lawyer or a bank, are generally passed to a trustworthy third party. If the other party is confident that their contract conditions have been met by both sides of the transaction, they fork over the money and agreements likewise and appropriately. Now, without involving a third party, a better technology-based smart contract could do this effectively and easily. A global education marketplace by Odem connects students, educators, professionals, and researchers who are interested in continuing their education.

23.2 Pay Degree Course for Pay

Bitcoin is approved by many, but until now, only the University of Nicosia in Cyprus, Greece, allows students to pay a tuition fee for a Master of Science degree in Digital Currencies; this makes sense and a new way of education toward digital transformation.

24 Conclusion

This chapter addressed several measures and secretively private control technology as a solution for blockchain technology for the business implementations and the attempts of Fujitsu to sell blockchain-based technology products and related services by presenting the degree to which Fujitsu is quite interested in the OSS project model. Blockchain technology, unlike a traditional centralized data processing system, incorporates information into a unique ledger, retaining continuity, while the organization is decentralized. Thus, we can say that it is a revolutionary technology for the introduction of minimum cost tamper resistance and high availability information management systems, contributing to the next generation of ICT. Upcoming projects and advancements will involve more detailed and precise efforts to expand business applications and their technology by developing various components of latest advancements with the better output and performance level.

References

1. Swan M (2015) Blockchain blue print for New Economy, O'Reilly Media, Inc. Sebastopol, CA
2. Kitchenham B, Charters S (2007) Guidelines for performing Systematic Literature Reviews in Software Engineering
3. Coinmarketcap (2016) Crypto-currency market capitalizations. https://coinmarketcap.com/. Accessed: 24 Mar 2016
4. Nakamoto S (2008) Bitcoin: a peer-to-peer electronic cash system. Consulted 1(2012):28
5. Kondor D, Pósfai M, Csabai I, Vattay G (2014) Do the rich get richer? An empirical analysis of the Bitcoin transaction network. PLoS One 9(2):e86197. pmid: 24505257
6. Herrera-Joancomart J (2015) Research and challenges on bitcoin anonymity. In: Garcia-Alfaro J, Herrera-Joancomart J, Lupu E, Posegga J, Aldini A, Martinelli F et al (eds) Data privacy management, autonomous spontaneous security, and security assurance, vol. 8872 of lecture notes in computer science. Springer International Publishing, pp 3–16. Available from: https://doi.org/10.1007/978-3-319-17016-9_1
7. Bitcoincharts; 2016. Accessed: 24 Mar 2016. https://bitcoincharts.com
8. Housley R (2004) Public Key Infrastructure (PKI). Wiley. Available from: https://doi.org/10.1002/047148296X.tie149
9. Double-spending; 2016. Accessed: 24 Mar 2016. https://en.bitcoin.it/wiki/Double-spending
10. Bitcoinwiki; 2015. Accessed: 24 Mar 2016. https://en.bitcoin.it
11. Antonopoulos AM (2014) Mastering Bitcoin: unlocking digital cryptocurrencies. O'Reilly Media, Inc.
12. Proof-of-Stake; 2016. Accessed: 24 Mar 2016. https://en.bitcoin.it/wiki/Proof_of_Stake
13. Borenstein J (2015) A risk-based view of why banks are experimenting with Bitcoin and the blockchain. Spotlight on Risk Technology. N.p. 18 Sept 2015. Web. 03 May 2016
14. Barski C, Wilmer C (2014) The blockchain lottery: how miners are rewarded - CoinDesk. CoinDesk RSS. CoinDesk, 23 Nov. 2014. Web. 03 May 2016
15. Wild J, Arnold M, Stafford P (2015) Technology: banks seek the key to blockchain - FT.com. Financial Times. N.p. 1 Nov. 2015. Web. 03 May 2016

An Approach to Ensure High-Availability Deployment of IoT Devices

Abishaik Mohan, Balaji Seetharaman, and P. Janarthanan

1 Introduction

The rapid evolution and integration of smart devices catalyzed by the Internet of Things (IoT) [1] have caused wireless networks to expand in an unprecedented way. According to a recent research survey [2], by 2025, the number of connected devices is estimated to exceed 50 billion. It is important to note that these estimates are evaluated based on what is true today and not on a wide variety of external advancements in other technologies. This digital transformation has caused demands for information systems that are error-prone and inclined to sag when scaled to trillions of devices leading to disastrous failures. Smart connectivity of things to existing networks and computing in the context using highly available network resources that are resilient to failures is an essential part of IoT. Consequently, assuring users of ubiquitous communication quality is the ultimate goal for the optimal Access Point (AP) deployment. To forecast the network conditions, it is significant to understand the model and behavior of network variables. A real-time evaluation with appropriate space-related parameters would allow us to predict the exact performance and network issues in a given point of time. The latest industrial standards support mobile data offloading [3], showing how the Wireless LAN (WLAN) technology is being merged with cellular technology [4]. New networking technology such as 5G will lead to growing wireless networks. Besides, improving the current infrastructure is time-consuming and expensive. Wireless LAN infrastructure in 5G networks is projected to play an important part in shaping how services are delivered to businesses and consumers.

A. Mohan · B. Seetharaman (✉)
Extreme Networks, Chennai, India

P. Janarthanan
Sri Venkateswara College of Engineering, Sriperumbudur, India
e-mail: janap@svce.ac.in

© Springer Nature Switzerland AG 2022
P. Raj et al. (eds.), *Blockchain, Artificial Intelligence, and the Internet of Things*,
EAI/Springer Innovations in Communication and Computing,
https://doi.org/10.1007/978-3-030-77637-4_5

The AP coverage rate, the High-Availability (HA) deployment phase, and the cost impact of the AP deployment are taken into account in this study. The static deployments of network resources [5], though optimal, don't ensure continuous operational efficiency when one of the critical resources goes down since it is not autonomous or adaptive. Though backup or stand-by resources are deployed as a substitute, they are poorly utilized. For instance, only when an active Access Point (AP) goes down, the idle access point becomes active and handles the load of wireless clients of the failed AP. The security of these connected devices as well as of the network is another vital factor to be considered, especially during the downtime [6]. Hence, an automated system that could predict the resource failure by analyzing large volumes of traffic patterns and adapt accordingly in a secure way would be highly beneficial.

The Multichassis Link Aggregation Group (MC-LAG) is a system that ensures additional redundancy to the LAG's normal link-level redundancy [7]. LAG can be applied in two distinct ways: LAG N and LAG N+N. LAG N is the worker-standby mode of load sharing of LAG and LAG N+N. It dynamically distributes and balances traffic across working links in LAG and maximizes the usage of the group while Ethernet links are down or up, offering greater intensity and efficiency [8]. Full implementation of the LACP protocol supports distinct LAG subgroups for a particular kind of resilience between two nodes. For LAG N+N, the worker's ties as a group would fail if one or more links in the working group fail [9]. MC-MLAG cluster nodes negotiate to ensure that automatic switchovers (failovers) are synchronized to guarantee HA [10].

Network Login is a measure of network security that authenticates users based on a web-based process, a MAC-based process, or as depicted in IEEE 802.1X. MAC-based authentication is used to authenticate systems based on their MAC addresses providing an additional authentication layer for smart devices [11]. For instance, if clients are permitted network access via station A, a MAC-based method is the one used for authenticating station A. Clients need to authenticate by other methods based on the appropriate network rights. MAC-based authentication can also be used as an external protection layer to authenticate Wi-Fi phones to deter other users from accessing the network using what is typically an unsecured SSID [12]. This, therefore, helps control the entry of user packets into the network by allowing only authenticated clients to reconnect to another AP in the case of an unplanned downtime [13]. Such architecture of a highly available system is designed and elaborated to ensure continuous efficiency in the case of a sudden downtime. In recent years, there have been interesting studies in the context of Wireless LAN AP Deployment [14], High-Availability Deployment [15–17], and Zero-Downtime deployment [18].

The rest of the sections are as follows. In Sect. 2, in comparison to our work, we discuss the related work. We provide a brief insight into the availability system in Sect. 3 and explain the MC-LAG components. The proposed architecture of ensuring zero downtime deployment based on High-Availability infrastructure is described in Sect. 4. In Sect. 5, we represent our system's experimental assessment and represent the findings. We conclude the paper with some future enhancements and discussions in Sect. 6.

2 Related Work

A couple of examples include the study of cloud computing architecture and how to orchestrate virtual machines (VM) in the cloud and the framework required to host such applications to ensure HA [19, 20]. The IoT architecture continues to face many significant hurdles, mainly among them being security, as the environment incorporates various parameters. Common security mechanisms such as lightweight encryption, stable protocols, and privacy assurance are inadequate. L. Perkov et al. and E. Braastad et al. focused on the same by using open source virtualization technologies and building applications with resources such as Heartbeat or Pacemaker [21]. Both methods use some kind of availability cluster to build network-level redundancy and then use software to route and track traffic between them. P. Endo et al. mentioned the five-nines or 99.999% availability as the requirement and goal for what should be considered as high reliability [22]. It is experimented by the amount of time available for a given period, usually over a whole year [23]. However, no work was found on applying these methods to a high-availability architecture.

The technology behind this failover strategy is the arbitrator service, which improves the user's functionality by tracking the communication between a user and the internal components. This tracking is carried out regularly by transmitting messages to an external body or a platform. If the primary component is down, all user requests must be forwarded to the secondary component, and the primary node must be restored or reset by the administrator manually. During this time, the whole device would no longer be usable if any further error occurs. In comparison, the primary/secondary solution does not address the scalability issue since only one device instance is usable at any time. There are also a few other options when it comes to Wireless Local Area Network (WLAN). For the heavy traffic load expected to occur in the urban areas, with extensive Wi-Fi usages, manual strategies were suggested to prevent downtimes. E. Bulut et al. [24] proposed to deploy WLAN APs in places with high data demand and maximize the total offload traffic from the cellular network. Subsequently, this leads to the assumption that the downtime event may be found over the same period.

In general, there would be a similarity of usage over time and space for neighboring APs, so it is necessary to take into account network utilization. Wang et al. [25] attempted to deploy WLAN APs for Mobile Users (MUs) to provide pervasive connectivity of high quality by considering the coverage rate, budget, and power ratio. Both refused to acknowledge that unloading traffic from the wireless network of Mobile Network Operators (MNOs) could also decrease the income of MNOs [26]. The problem of deploying WLAN APs that can provide MUs with wireless services was studied in Ref. [27], and two issues were resolved to optimize the coverage of the continuous MUs and minimize the cost of deployment of wireless LAN APs. It, however, failed to solve the profit–loss issue.

This research paper outperforms the currently proposed methods and ensures continuous operational efficiency without comprising the security and at the same time effectively utilizing the deployed resources. It recommends an approach to the use of MC-LAG architecture that is used specifically to achieve High Availability. During an unplanned downtime in wireless networks, an automated process is triggered, which extracts the details of connected devices to the attached AP and reconnects it to another AP securely within a short time. Experimental results show that the proposed approach is effective in real-time scenarios.

3   High-Availability Architecture

The core elements of the HA architecture are presented in this section, and the functionalities of its components are explained. The formulation of the MC-LAG architecture and its implementation in the backhaul network is also elaborated in the process.

3.1   Network Reliability

It is difficult to measure network output efficiency since the state space increases by the scale of the network. The efficiency of the network output is quantified as the likelihood that performance metrics stay within the predicted limits of a given traffic flow. In this chapter, the key measurement is a delay since customers generally care most about this performance, which is an acceptable metric for realistic communication networks.

The elementary performance measures of network deployment are its utilization and connectivity, even though it is secondary to data rate, throughput, and channel utilization. Availability refers to a device capable of delivering continuous operational service under standard conditions. In wireless deployments, high redundancy is usually provided using Wireless Controllers, as depicted in Fig. 1.

Wireless networks run smoothly under a wide variety of strict operating requirements. As far as wireless networks are concerned, the problems of co-channel interference, signal failure or fading, the presence of barriers, and the transmission capability of the network should be taken into consideration. Interference is also a huge concern. Caused by several sources, the wireless network can easily be influenced by interference because wireless infrastructure uses unlicensed bands. Interference occurs when the reception of a signal from a node in the network is distorted by another signal from other sources.

Another challenge is the time-sensitivity criterion, that is, the time-critical data should almost reach in real-time. Furthermore, given that connection and node failures are not unusual to Wireless Sensor Networks (WSNs), it is a difficult task to ensure timely reporting of results. Data loss occurs if proper procedures for aggregating data are not adopted. The aggregated data obtained by the primary node is most relevant, though, since this information is the basis for taking effective action in an emergency. Since power limit sensors typically do not transmit data from

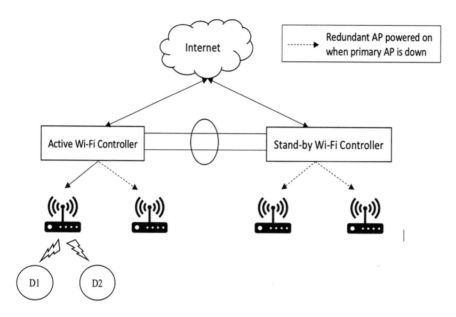

Fig. 1 Conventional access point deployment

"events," there may be a discrepancy in what the primary node collects as aggregated information and what occurs. Therefore, identifying this also becomes a challenge.

Mean Time Between Failure (MTBF) is a proof of concept used to indicate the device's failures per million hours. Mean Time to Repair (MTTR) is the average time required to patch a failed component and make it functional. Availability can be evaluated as follows:

$$\text{Availability} = \text{MTBF} / (\text{MTBF} + \text{MTRR}) \tag{1}$$

The longer it is possible to operate a system on average until a breakdown happens, the greater its reliability. Thus, since a device has an uptime that is considerably greater than the amount of downtime, high availability is guaranteed. MTTR, which is the total time it takes to restore and restart daily service after a system breakdown, tests maintainability. The higher the system's maintainability is, the shorter the total maintenance period. If we describe the period between downtimes as D1 and duration between uptimes as D2, the availability A can be determined as follows:

$$A = \text{D1} / (\text{D1} + \text{D2}) \tag{2}$$

3.2   Multicast—Link Aggregation Group

An MC-LAG is a type of LAG that centralizes constituent ports on separate chassis, which primarily serves as a reliable load functionality to increase bandwidth and provide redundancy in an emergent breakdown of one of the components. In our implementation, the backhaul network is configured as a cross-connect MC-LAG, as depicted in Fig. 2.

It assures greater stability, fault tolerance, and operation uninterrupted with a meshed path existing between the source and destination. The MC-LAG allows two MC-LAG peers to form a logical LAG interface to a client computer. It provides redundancy and load balancing without running the Spanning Tree Protocol (STP) between two MC-LAG peers, multihome support, and a Layer 2 loop-free network. It is a multiprotocol HA solution. An MC-LAG client, like a server, has one or more physical links on one end of an MC-LAG in a link category. No MC-LAG must be configured for this client device, and two MC-LAG peers are on the other side of MC-LAG.

For example, a standard layer-2 topology could mean several redundant layer-2 connections between two endpoints that are not error-disabled by including the multichassis link aggregation group (MC-LAG) configurations. Another representation of a layer-2 topology may, however, represent fewer endpoints than are currently present to give a perception that a series of link-layer connections exist on the same endpoint.

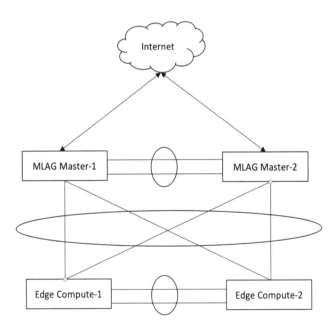

Fig. 2 Cross-connect MC-MLAG

Each peer has a physical link to a single client device, and the peers use Interchassis Control Protocol (ICCP) to share and organize control information to ensure proper routing of data traffic. If any of the switches fail, Link Aggregation Control Protocol (LACP), which detects and protects the network from any misconfiguration, reconfigures the path in as little as few seconds. In addition to MTTR and MTBF, convergence time is a key factor in determining the performance of the communication network. Networks converging faster are very stable.

Enterprises that have low fault-tolerant deployment infrastructure could gradually replace it with an autonomous and adaptable architecture. Traffic pattern analysis could be taken advantage of, which will provide intelligence to the deployed resources regarding the appropriate steps to be taken during unstable conditions. The above deployment results in the ease of management of the network devices. Hence, the FCAPS (Fault, Configuration, Accounting, Performance, and Security) guidelines can be achieved with minimal effort.

3.3   Security

A detailed and accurate representation of the network topology used for security analysis needs specific network endpoints. User recognition by interface attributes, such as associated addresses, normally occurs when identity assurance attributes such as a Trusted Platform Module (TPM) are rooted in hardware. An inadequate number of policies can be enforced to ensure safety protocols, such as Trustworthy Network Link, if only the topology is known. The consistency of a network—particularly in terms of completeness and precision—has a strong effect on the quality and efficiency of the security present.

If no authentication or a weak encryption is used, WLANs are vulnerable to attack. Besides, an intrusion by Man-in-the-Middle (MiM) becomes a big problem for WLAN users and owners. Various forms of attacks have been documented and widely researched by scholars around the world. Though efforts are increasingly being made in the later iterations to resolve security vulnerabilities, WLAN security remains a concern. Despite the changes made for WLANs, the wireless network still demands further safety.

With the connectivity among devices increasing exponentially, the current method of ensuring network security is not optimal, especially for the IoT devices that have constrained resources. Consequently, security must be higher. Net Login is a network-level authentication that confirms user access to services. When a user attempts to log in to the network, they indicate their identity with their Media Access Control (MAC) address learned on a particular port, which is called MAC-based auth.

A system then cross-checks the learned MAC with a predefined list of authorized MACs to ensure they are granted access to the network. It is especially vital to ensure that no compromise occurs with the data or on the end device. Kernel-based Virtual Machine (KVM) is a full virtualization solution where the insight interfaces are generally several dedicated high-speed links. They are mounted as Ethernet interfaces of an isolated Ubuntu VM, shortly called Third-Party VM (TPVM). This physical separation allows secure transfer of voluminous traffic with a minimal delay while not causing performance degradation of the control or data plane. Together, this makes it possible to run tools and services directly on the device without deploying additional infrastructure and at a much cheaper cost.

4   Proposed Methodology

In this section, the proposed High-Availability deployment infrastructure, as shown in Fig. 3, is discussed.

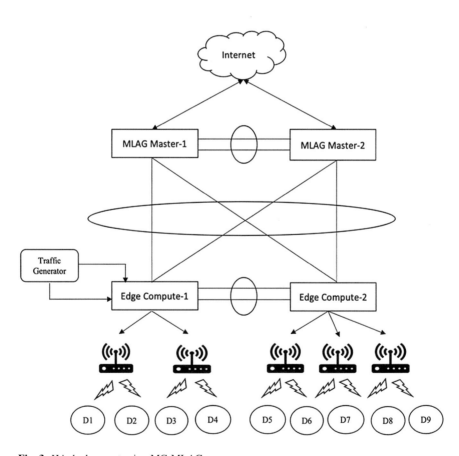

Fig. 3 HA deployment using MC-MLAG

4.1 Initial Deployment

The wireless clients are connected to AP1 and AP2, which are in turn connected to PoE (Power over Ethernet) edge switches, E1 and E2, which are supported by the backhaul network M1 and M2 that are implemented using cross-connect MC-MLAG. The APs are authenticated using MAC-based Network Login, and the details are configured in the edge switches. Similarly, in each AP, the end devices are authenticated with their MAC entries preprogrammed using MAC auth access-list in the APs. Before the completion of deployment, a thumb rule is to ensure that the number of end devices associated with a particular AP is always half the maximum limit; if the maximum limit is "n," the number of wireless clients associated with an AP is "$n/2$" before the failover scenario occurs.

In case an unplanned downtime such as a network disconnection or a power outage occurs, Convergence Time (CT) plays a crucial role in determining the number of packets lost during the switchover. It is calculated as follows:

$$CT = \left(\text{Transmitted} - \text{Received}\right)\text{Cumulative Frames} / \text{Transmitted Frame Rate} \tag{3}$$

As each route is switched over, traffic for that route starts appearing on the backup link. It is not until the last route has been switched that convergence can be completed. The SSID (Service Set IDentifier) credentials of AP1 and AP2 are initialized in the wireless clients and are then set to station mode.

The end devices then scan the network at a time to determine the available APs, which are under the coverage and connect to the nearby AP. If no active AP is present, the end devices keep scanning the network periodically through a watchdog reset.

A successful connection is established only if the end devices achieve authentication through Net Login. The initial sequence of steps depicting the connection establishment of an end device with an AP is illustrated in Fig. 4. To ensure that the APs are under the signal coverage, they are deployed optimally using a region-based or an analysis-based heuristic approach that would guarantee redundant APs for the end devices. After the initial deployment is complete, a triggered process extracts the details of IP reachability from all the deployed devices and stores them in the edge switches and the backhaul network.

4.2 Ensuring HA During Failover

Downtime can be predicted in certain cases by analyzing traffic patterns of past data when a particular resource had gone down especially during the peak time where the frequency of access would be high. In such cases, High Availability can be

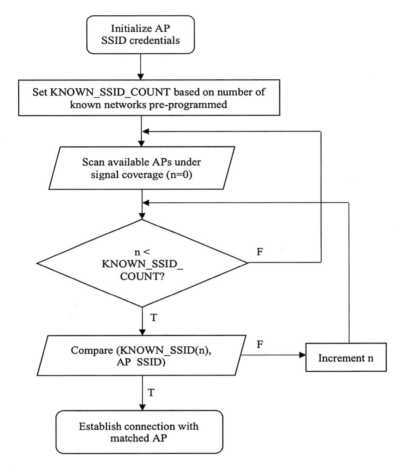

Fig. 4 Initial connection establishment

ensured by taking suitable steps to move user traffic to the redundant device, which is also under the coverage of end devices.

This redundancy was ensured during the initial deployment. When a link flap on either of the MLAG switches or the edge switches is detected, a log event is gener- ated, which triggers the code to be run in TPVM of the redundant switch. Appropriate API (Application Programming Interface) calls are made, which configures the MAC addresses of the wireless clients to the redundant AP and checks if the recon- nection was successful. Since the initial count of connected devices was "$n/2$," the count is now increased to "n" during a switchover.

This ensures that the deployed resources are effectively utilized, unlike having a stand-by AP that is partially utilized. Since the end device MACs are now prepro- grammed on the redundant AP, they can securely authenticate again, similar to the scenario when they were initially deployed. The above topology uses a symmetrical scenario. The same principle could be applied in the case of an asymmetrical

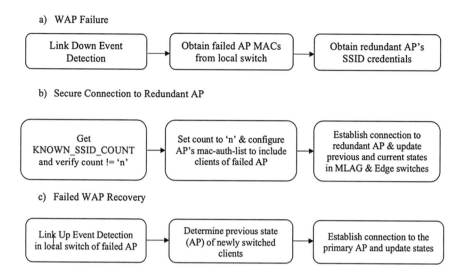

Fig. 5 Sequential diagram

scenario and extended to span a huge network. We represent the various stages occurring during a failover in the sequential diagram depicted in Fig. 5. Whenever the AP or edge switch recovers from the failure, the system could be restored to normalcy by reconnecting the end devices to the previous AP and resetting the limits to "$n/2$." This could be done either manually after an investigation by network admin or by automatically keeping track of the switched states.

5 Experimental Results

An experimental analysis is done by applying the proposed architecture to confirm whether High Availability is accomplished. A series of iterations were performed using a few IoT devices to determine the average failover and convergence times, and the results were compared with the conventional deployment scenarios.

In our experiment, we used NodeMCU (Node MicroController Unit) as the IoT device, which is an open-source Lua-based firmware for ESP8266 Wi-Fi SOC (System-on-a-Chip). We experimented with verifying if the proposed architecture could produce optimal failover and convergence times both during planned and unplanned downtime. In the case of planned downtime, we increased the frequency of operations on the AP, resulting in stress, to more than the maximum value that was obtained while testing the infrastructure.

On detecting extremely unusual peak load, the edge switch was successfully able to reconnect the end devices securely to the nearby AP without disrupting the service. In the case of unplanned downtime, we simulated power outage and network disconnection in the AP and the edge switches in a series of three iterations. On

Table 1 Failover and convergence time measurements

Iterations	Failover time (s)	Convergence time (ms)
1.	7.89	53.70
2.	7.88	54.23
3.	7.85	55.23
Average	7.87	54.33

```
Setup done                          22:11:59.311 -> 1 network(s)          22:12:47.470 -> Ping statistics for
                                    found                                8.8.8.8:
22:11:00.725 -> scan start          22:11:59.311 -> Airtel-Hotspot-      22:12:47.470 ->    Packets: Sent
22:11:02.899 -> scan done           07E0                                 = 50, Received = 42, Lost = 8
22:11:02.899 -> 2 network(s)        22:11:59.311 -> -52                  (10.00% loss),
found                               22:11:59.311 -> dBm :                22:12:47.470 -> Approximate
22:11:02.899 -> Vodafone            22:11:59.311 -> 96                   round trip times in milli-
22:11:02.899 -> -33                 22:11:59.311 >                       seconds:22:12:47.470 ->
22:11:02.899 -> dBm :               22:11:59.311 > Connecting to         Minimum = 44ms, Maximum =
22:11:02.899 -> 100                 Airtel-Hotspot-07E0                  72ms, Average = 54.33ms
22:11:02.899 ->                     22:11:59.835 -> Request timed        22:12:47.470 -> Destination host
22:11:02.899 -> Connecting to       out.                                 data:
Vodafone                            22:12:00.335 -> Request timed        22:12:47.470 ->    IP address:
22:11:03.454 -> ...........         out.                                 8.8.8.8
22:11:09.087 -> WiFi connected,     22:12:04.951 -> WiFi connected,
your IP address is                  your IP address is
22:11:09.087 -> 192.168.1.197       22:12:04.951 -> 192.168.1.102
22:11:17.086 ->                     22:12:05.455 -> Request timed
22:11:17.086 -> waiting for the     out.
failover event                      22:12:05.490 -> Reply from
22:11:22.086 -> waiting for the     8.8.8.8: bytes=32 time=50ms
failover event                      TTL=114
22:11:27.080 -> waiting for the     22:12:06.534 -> Reply from
failover event                      8.8.8.8: bytes=32 time=54ms
22:11:57.102 -> Pinging google IP   TTL=114
8.8.8.8                             22:12:07.544 -> Reply from
22:11:57.102 -> scan start          8.8.8.8: bytes=32 time=55ms
22:11:58.085 -> Request timed       TTL=114
out.                                22:12:08.501 -> Reply from
22:11:59.118 -> Request timed       8.8.8.8: bytes=32 time=53ms
out.                                TTL=114
22:11:59.311 -> scan done
```

Fig. 6 Serial logs obtained in NodeMCU

average, we achieved a good failover time of 7.87 seconds and a convergence time of 54.33 milliseconds, as can be seen from Table 1.

The serial logs obtained during the experiment can be seen in Fig. 6. Hence, the static deployments could be gradually changed to adapt to such scenarios wherein optimal results could be achieved. The whole process is automated, and the user has the option of either intervening manually to restore the system to normalcy or issue an automatic restore that tracks and updates the states automatically. ESP8266, being an IoT device, operates at 80 MHz, and this is the reason for its quick switchover. If the operating frequency is increased, the switchover time could be lowered much further.

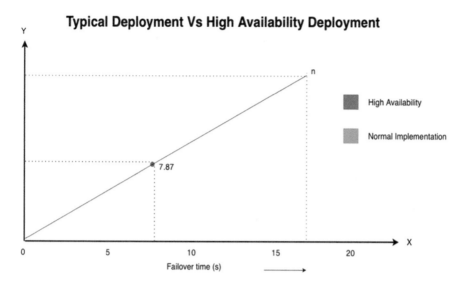

Fig. 7 Failover time measurement

As shown in Fig. 7, the failover time obtained is low using our architecture when compared with the typical implementation. Therefore, the proposed automated deployment can be utilized without any compromise in user-level and network-level security. When a huge network spanning multiple nodes is considered, the performance optimization obtained is also comparatively high since all available resources are utilized efficiently.

6 Conclusion

This chapter discusses the issue of ensuring constant operational efficiency using an automated model of High-Availability deployment without compromising the security of end devices. It proposes an approach that is efficient compared with the typical HA deployment scenarios. Typically, resource-constrained devices lack security despite the ongoing efforts to bridge the gap. The discussed methodology helps resolve this problem and its associated connectivity issues that are typically expected in a WLAN infrastructure. Experimental results showed that the architecture produces optimal results during a planned or an unplanned downtime with a failover time of 7.87 seconds, as shown in Fig. 5.

This work can be extended by unifying the monitoring and management of such resource-constrained devices in a particular network through the cloud. In the future, optimizing the network may be worth trying to obtain better results. To improve the convergence time, end devices having high operational frequency could be used to reinforce the network more. We would also continue to expand this approach by utilizing secure boot, which verifies the authenticity of the firmware image running in IoT devices so that the intruders cannot attack the firmware and replace it with malware during any active operation. ESP8266 would be an ideal platform since it has the capabilities to support cryptographic algorithms and a Trusted Platform Module (TPM) to store the keys securely.

References

1. Miorandi D, Sicari S, Pellegrini FD, Chlamtac I (2012) Internet of things: vision, applications and research challenges. Ad Hoc Netw 10:1497–1516
2. Al-Fuqaha A, Guizani M, Mohammadi M, Aledhari M, Ayyash M (2015) Internet of things: a survey on enabling technologies, protocols, and applications. IEEE Commun Surv Tutorials 17(4):2347–2376
3. Alcatel and British Telecommunications (2012) Wi-Fi roaming building on andsf and hotspot2.0. White Paper
4. Cisco Systems (2011) The future of hotspots: making Wi-Fi as secure and easy to use as cellular. White Paper
5. Frangoudis PA, Polyzos GC (2010) Report-based topology discovery schemes for centrally-managed Wi-Fi deployments. In: Next generation internet (NGI), 6th EURO-NF conference, pp 1–8
6. Roman R, Lopez J, Najera P (2011) A cross-layer approach for integrating security mechanisms in sensor networks architectures. Wirel Commun Mob Comput 11(2):267–276
7. Wang Y, Ren Y, Meng Y, Bai J (2017) Research on performance of three-layer MG-OXC system based on MLAG and OCDM. In: Proceedings of SPIE 10464, AOPC, 24 October 2017
8. IEEE 802.1AX – Link Aggregation, IEEE, 2008
9. Farkas J, Antal C, Westberg L, Paradisi A, Tronco T, de Oliveira V (2011) Fast failure handling in ethernet networks. IEEE Int Conf Commun 2:841–846
10. Irawati LD, Hariyani YS, Hadiyoso S (2017) Link aggregation control protocol on software defined network. Int J Electr Comput Eng 7(5):2706
11. Xiao Y et al (2010) Security services and enhancements in the IEEE 802.15.4 wireless sensor networks. In: Proceedings of the IEEE Global Telecommunications Conference (GLOBECOM '05), November 2010
12. Garcia-Morchon O et al (2011) Security considerations in the IP-based Internet of Things. IETF. http://tools.ietf.org/html/draft-garcia-core-security, March 2011
13. Lu B, Pooch UW (2005) A lightweight authentication protocol for mobile ad hoc networks. Int J Inf Technol 11(2):119–135
14. Cheung MH, Huang J (2015) DAWN: delay-aware Wi-Fi offloading and network selection. IEEE J Sel Areas Commun 33(6):1214–1223
15. Park S, et al (2013) An analysis of replication enhancement for a high availability cluster. J Inf Process Syst 9(2):205–216
16. Brenner S, Garbers B, Kapitza R (2014) Adaptive and scalable high availability for infrastructure clouds. Distributed Applications and Interoperable Systems. Springer, Berlin, Heidelberg, Systems, pp 16–30
17. Kim H, Feamster N (2013) Improving network management with software-defined networking. Communications Magazine, IEEE 51(2):114–119
18. Cleveland F (2012) Enhancing the reliability and security of the information infrastructure used to manage the power system. In: IEEE PES general meeting, Tampa, 24–28 June 2012
19. Moreno-Vozmediano R et al (2018) Orchestrating the deployment of high availability services on multi-zone and multi-cloud scenarios. Int J Grid Utility Computing 16(1):39–53

20. Cziva R, Stapleton D, Tso FP, Pezaros DP (2014) SDN-based virtual machine management for cloud data centers. In: IEEE 3rd international conference on cloud networking (CloudNet), pp 388–394
21. Benz K, Bohnert TM (2014) Impact of pacemaker failover configuration on mean time to recovery for small cloud clusters. In: Cloud computing (CLOUD), IEEE 7th international conference on, pp 384–391
22. Park J, Jeong J, Jeong H, Liang C-JM, Ko J (2014) Improving the packet delivery performance for concurrent packet transmissions in WSNs. IEEE Commun Lett 18(1):58–61
23. Nai W, Zhang F, Yu Y, Dong D (2015) Reliability evaluation for key equipment of high-speed maglev operation and control system in uncertainty environment based on fuzzy fault tree analysis. Proceedings of MAPE Conference, Shanghai
24. Bulut E, Szymanski BK (2012) Wi-Fi access point deployment for efficient mobile data offloading. Proceedings of 1st ACM International Workshop on Practical Issues and Applications in Next Generation Wireless Networks (PINGEN), Turkey, pp 45–50
25. Wang CS, Kao LF (2012) The optimal deployment of Wi-Fi wireless access points using the genetic algorithm. In: Proceedings 6th International Conference on Genetic and Evolutionary Computing (IGCEC 2012), Kitakyushu, Japan, pp 542–545
26. Zhang C, Gu B, Yamori K, Xu S, Tanaka Y (2013) Duopoly competition in time-dependent pricing for improving the revenue of network service providers. IEICE Trans Commun E96-B(12):2964–2975
27. Wang M, et al (2016) Cellular machine-type communications: physical challenges and solutions. IEEE Wireless Commun 23(2):126–135

Blockchain-Based IoT Architecture for Software-Defined Networking

P. Manju Bala, S. Usharani, T. Ananth Kumar, R. Rajmohan, and M. Pavithra

1 Introduction

The Internet of things (IoT) relates to the millions of physical appliances around the globe that are currently interconnected to the Internet, not only capturing and exchanging data but having users do things such as build an alert in the case of a smoke detector or make a smart light dim, in the case of a dimmer. We were shocked to learn that smart cities and smart nations are evolving right as the Internet of things (a word referring to a vast amount of microchips and sensors being used in all kinds of devices) and blockchain technology (the application of cryptography in distributed ledgers) are taking off. According to the specifications of these innovations, for example, the mode of implementation and reception of data sources has also been modified. IoT tends to impact all aspects of human life, including telecommunications, with the increasing development of IoT that is expected to improve by approximately 20–50 billion worth of industries by 2020 [1]. IoT faces challenges such as the absence of a centralized unit, system heterogeneous nature, full attack, and resemblance [2]. Protection and energy usage are the most important challenges among the IoT domain names [3]. For instance, due to complexity in energy supplies and storage, IoT devices have constraints and thus lead to possible inefficiencies in coordination and execution of security systems [4]. Cloud computing has currently been found to be a feasible design to solve various difficulties related with the restrictions on IoT resources [5]. But, there are unsolved risks that occur when adapting security for the Internet of things. Here in the real world, we find our current technology needs to follow functions relating to safety, accessibility, and high security as well as power consumption for IoT devices to communicate with each other. Integrity is a mechanism that ensures that the NSA can monitor

P. Manju Bala (✉) · S. Usharani · T. Ananth Kumar · R. Rajmohan · M. Pavithra
Department of Computer Science and Engineering, IFET College of Engineering,
Villupuram, Tamilnadu, India

© Springer Nature Switzerland AG 2022
P. Raj et al. (eds.), *Blockchain, Artificial Intelligence, and the Internet of Things*,
EAI/Springer Innovations in Communication and Computing,
https://doi.org/10.1007/978-3-030-77637-4_6

only approved users. The message is sent to the recipient in a clear, meaningful manner and can be adjusted without detection of the same message being sent to the same recipient. Think of usability as making sure that the system is usable at all times. However, think of the network of connected devices, IoT devices as being connected with a minimal machine-to-machine contact. Thus, accessibility is more important than computing power and energy since these devices typically have restricted horsepower. Reducing the amount of energy used for data transmission on an IoT network is an essential factor, but it must be balanced with energy conservation and safety. This is a new architecture of IoT servers that requires the creation of new aspects. Our current approach for IoT still causes power wastage for physical IoT and data services running in the cloud, and we have to build an architecture with the requisite functionality to fulfill the actual need for security as well as energy utilization for IoT being on separate layers. Through using software-defined networking (SDN) and blockchain technology, we can build an infrastructure that can organize groups of computers in a coordinated manner.

SDN stands for software-defined networking, which starts from a base of pure computing along with abstraction layers separating the control plane from the data plane. It is connected to a rule-based switch (or to a switch like OpenFlow), which routes packets and, by setting it to certain rules, establishes governance, configuration, and malfunction in each individual networking device. Features such as absolute influence on the device, announcement and remote and intelligent control, high versatility, and completely programmable can be accomplished by the SDN controller. In addition, the SDN organizer enables the capability to execute centralized and network security resources such as safety, scheduling, power consumption, and control of frequency range and can block malicious entry to the network constraints [6]. When the IoT comes around, software developers would need care in problem-solving, solution finding, and programming, so SDN should handle it all. Since artifacts are often generated and utilized via IoT program code that is executed on thin-client mobile devices, the emerging network infrastructure needs to be built and brought to a "point of readiness" via SDN controllers that are more easily controllable and managed. SDN features can be shared with the IoT to improve the presentation of the network. One of the main problems of the IoT network is that it does not have any centralized unit, which can be solved by utilizing the SDN controller to issue one such centralized controller that communicates with IoT devices. Among the SDN's biggest problems today is to improve its position through steady growth and development. When it comes to file sharing, the distributed file system (SDN) is more accurate and secure than the existing blockchain. The security of the life of blockchain can be used through SDN controllers for IoT devices that connect to the cloud to safeguard the accessibility and confidentiality of resources from insecure users regarding energy consumption. Blockchain is a distributed ledger that facilitates the validation of transactions through Bitcoin. Blockchain was used to develop a wide range of computer networks in addition to big data applications like health care, e-government, and decentralized cloud storage next to financial transactions. A blockchain is a cryptographically connected collection of transactions that include data that often use distributed and scalable peer-to-peer network infrastructure, rather than centralized management. Doing things the correct way is ensuring that there are no intermediate steps. This is a system without any intermediaries. Using the SDN (Stretched Device Network) capability to develop a robust and energy-efficient infrastructure, we overcome the difficulties and shortcomings of the IoT with blockchain technology. The aim of the P2P technology is to provide a secure means of communication for the main issue of security, safety, quality, performance, and solutions through problems (single point of failure). Generally speaking, when we store data, they are stored on one server. When sending data out, they are located on another. When combining blockchain with SDN, we can store data on a blockchain and send data out on an SDN. Based on this, a new blockchain-based IoT architecture for SDN is proposed. The suggested blockchain layer framework permits a decentralized P2P system where unreliable entities may communicate in a provable way with each other in each cluster without a trusted third party (SDN cluster). In each SDN field, this enables safe interaction among IoT devices. It is, however, recognized that blockchains, like bandwidth overload and latency, could exacerbate the computational elements; therefore, it may not be sufficient for IoT devices. Our suggested architecture offers a modified and IoT blockchain to solve this problem, which basically removes the latency of the conventional blockchain through an effective authentication system with shared confidence in each cluster by the SDN. More specifically, to accomplish data storage using the Internet of things, each cluster possesses an internal storage database that the network layer of the platform keeps in control via SDN. The inclusion of malicious nodes is avoided to maximize POW, and appropriate routing protocols are used in the suggested IoT system architecture in the SDN controller. To maximize energy efficiency and improve safety, we use a cluster structure. Cloud processing (CP) is accepted by cloud computing as one of the main tools offered [7]. However, it is defined as a platform for cloud services, where information is maintained on a web server and can be accessed from the repository via the Internet. In addition, it is categorized into four types, including private, public, private, and hybrid cloud computing, operating under four levels, such as connection, functionality, governance, and memory. Cloud storage enables both consumers and commercial organizations to store data in the cloud that can face security issues such as data safety, privacy, and quality of data.

It is possible to utilize another new SDN platform as a solution to such issues. The aim of the SDN is to transform the network by splitting the control plane that handles the network from the data plane where data flow so that the system can be handled by an external controller. In addition, network virtualization comes with advantages, such as dynamically flipping the network upward, fine-tuning for multiple instances of system use, and implementing security measures on each individual system. Thus, while attackers are able to obtain the database, they are limited to a specific portion of the network that will limit their impact. But the cloud processing security problems cannot be solved by SDN technology itself because they have established DDoS attack problems themselves when SDN is combined with cloud processing. Blockchain, on the other hand, is another leading technology that describes a process to address the previously described problems. It is a chain block that contains data. Its basic feature is that it is very difficult to change once information is stored within a blockchain. Each block includes certain details, the block hash key, and the previous block and transactions section hash. The hash is essentially used with all its components to define the block, and it is special. Again, transaction information is firmly stored on the transaction part.

2 Related Works

Related work will review the associated investigation readings focusing on the advancement of SDN with blockchain and IoT technology. The exponential development of the Internet of things (IoT) has created new connected devices that may not be adequate for traditional distributed databases, protocols, and strategies to address unimaginable IoT challenges, predominantly those connected to safety and energy. The BC and SDN have involved the attention of investigators at the university because of their potential to provide technological solutions to current problems. DistBlockNet is a blockchain design with established patterns that uses a fixed blockchain to make sure that the concepts tables are solid and have self-protection in any devices that enforce them. This model is focused on SDN shared feature concepts for the generation of a plan focused on IoT system security and standardization concepts. Through modifying the flow rules columns using the blockchain, it gains from the ability to rapidly separate risks. However, the quantity of energy released by IoT devices and the limited resources do not take this architecture into account. In addition, problems with energy usage, in turn, may pose architectural security issues.

In the work in Ref. [8], a security policy on SDN is applied by Blockchain Security over SDN (BSS). In reality, it effectively executes file transfers via blockchain to the SDN. In this study, the use of the Ethereum framework and the open daylight system combined with the OpenStack framework demonstrates that files can be easily shared on the basis of distributed P2P between SDN users. This method has not been tested on the basis of the IoT standard since it merely discusses the problem of safe distribution and the areas of interoperability without taking into consideration the resource and energy constraints used. The authors in Ref. [9] have enhanced a blockchain-based solution for smart home IoT applications. Their suggested scalable architecture reduces the overheads of conventional blockchains. The architecture given by the shared consensus was used to decrease the verification time consumption. However, the suggested design does not take into consideration the drawbacks of IoT devices in terms of interoperability and resource utilization. The loss of routing between IoT devices in its hierarchical system can also further contribute to reliability issues and consumption of energy. A blockchain-based platform for IoT applications represents the architecture of a home automation example, similar to the one mentioned in the research report. The matter of connection is achieved by using two kinds of blockchains, and the authentication is done using a blockchain-based scheme. "Cooja is a peer-to-peer (P2P) scheme, which helps users to evaluate long-term quality success in joint projects." It was found that by the use of an error-correcting code, the different peers in a cluster support each other to reduce the amount of time for them to transmit data. The paper does not seriously consider the platform involved in IoT implementations, and the potential obstacles to that indicate limitations on how much they actually consider the possibilities of how IoT might be implemented. In Ref. [10], a classical blockchain-distributed cloud architecture was provided. This cloud node is situated at the edge of the network, and it includes distributed chip built to automate the entire network. Their architecture is made up of three layers, which contain cloud, fog, and applications, with cost-effective and blockchain on-request connectivity based on the distributed cloud environment. In our opinion, this design needs recognition of IoT concepts and obstacles, such as energy use, device-to-device connectivity, and IoT device resource constraints. In Ref. [11], a unique network architecture focused on the blockchain concept for smart cities has been proposed. To achieve performance, this fusion model is subdivided into two enterprise networks and edge networks. This hybrid model combines standardized functionality and the structure of a distributed network that uses a POW specification to preserve privacy and security. The design evaluation discusses features such as suspension, hashing rate, and block but does not take into account the concerns of IoT electricity usage and protection inquiry. In addition, in this hybrid network model, efficient implementation of edge nodes and allowing cache methods at the edge nodes seem to be difficult. Similarly, in Ref. [12], the authors suggested a block-VN-oriented car network model based on a smart city. A networked service architecture is called Block-VN as it is a framework through which several separate transmissions can communicate with each other via a special service. A network framework was proposed, which included vehicles that were linked via blockchain. We could also use a mesh net of blockchain vehicles that communicate with each other. This study examines the concepts of inter-vehicular network architecture in its tested situation while not discussing the criteria of assessment and IoT problems.

A stable, secure, scalable, and energy-efficient system called BC Confidence was built in using the traditional blockchain authentication scheme for processing, retrieval, and constraint of resources. The Wireless Sensor Network (WSN) was discovered to be part of the Internet of things in this investigation. Cyclic processing network (CPN) is a new form of processing network, which implements processes in a network of worker nodes. Each node is bound to a node called CPAN, which allows E-commerce websites to move and transact without using a bank and with complete confidence. Developing within the C language and the Ethereum blockchain is evidence that this is a profitable technique. Although it is true that blockchain distributes a distributed ledger of all "truth" to all of the IoT devices running on a specific blockchain, it doesn't significantly affect IoT resource management, particularly given that IoT resource management systems are already centralized. Since Ethereum is a simple implementation to use for smart contracts, it was used in their invention of it. The keys are regulated through RSA encryption in this work so that the public key is stored in an Ethereum contract and the private key controls it on various desktop and mobile systems. In the prototype analysis, it was found that networking constraints were not taken into account in the prototype system that is being developed, such as the IoT system that handles data. The authors addressed a broad, detailed SDN analysis for the identification and prevention of security problems [13]. The authors implemented a distributed controller framework [14] to address consistency, usability, low latency, and compatibility problems in a software-defined network, where clustering into the control layer would dramatically minimize CPU utilization and packet drop. In addition, all current SDN systems, such as Openstack and ONOS, proposed a specification. This architecture of an SDWSN model is used to recognize risks, problems, and potential solvents in the SDN and WSN approach, adopting two superior techniques called Software-Defined Networking (SDN) and Wireless Sensor Network (WSN). In addition, the authors [15] addressed the protection of numerous SDN technologies, and they proposed the most suitable security method based on security needs. In IoT [16], the Middlebox scheme and flowable ability limitations for SDN were suggested to manage transfer efficiency. On the other side, blockchain-based SDN technologies for the IoT system have already been implemented by several experts [17]. IoT design allows two big communication technology incentives, such as SDN and blockchain.

3 Proposed Blockchain-Based IoT Architecture

Distributed computing for IoT devices is being implemented in the proposed framework by chain connection using another form of centralized control system and a network use-driven SDN approach. The image below shows a high-level overview of the suggested framework for the SDN controllers that are connected to blockchain to allow IoT systems to communicate. Similar to the distributed networks used by Bitcoin, SDN controllers makes up a P2P network. The goal of this design is to induce a more stable communication pathway between IoT devices. The second goal is to reduce power consumption by the highest possible degree. Large networks normally do not operate in a healthy manner. To achieve efficient network functioning often needs to structure and prepare. Both need to occur in large networks. The reason for clustering is to promote this successful organizational outcome. This one area of the brain is like a "brain network cluster." The SDN controller acts as a cluster head—one that makes the network smoother and quicker, reducing latency and other variables. Each regional operator will have an IoT device developer who will be in charge of the execution. Suppose that the interconnected IoT devices are the other components of the SDN controllers in this architecture. The controller is the supervisor of each SDN cluster who handles compliance with requirements and policies of the cloud. They monitor the usage of IoT devices in each network. Another part of this architecture is cloud processing, which is used to handle communications and public blockchain between SDN controllers, and IoT systems stock their information in the cloud. One thing to remember is that the human factor is an important target for cyber-attacks and criminal abuse. The key concept of a blockchain is to protect the confidentiality of the data and to eliminate the single point of failure, which gives us the ability to use these blockchain features in a scattered collaboration between the SDN controllers in the hierarchical cluster system proposed (Fig. 1).

A stable and equivalent design in the proposed system is supported by the P2P network shared between the controller and the elimination of mediators for secure connectivity. Within the SDN realm, the P2P link between the IoT devices can be identified. We created a design where both public and private blockchains were used. Both blockchains, including ledgers, split the values of distributed database expertise associated with the decentralized peer-to-peer networks where each maintains a locally replicated version of a public digitally signed transactions joined on the database. Within the framework of our particular research, the heads in a cluster create and store a blockchain between one another.

Using the distributed architecture of a cloud network, a modern SDN controller and IoT gadgets can be integrated into the cluster network, and if a new block were

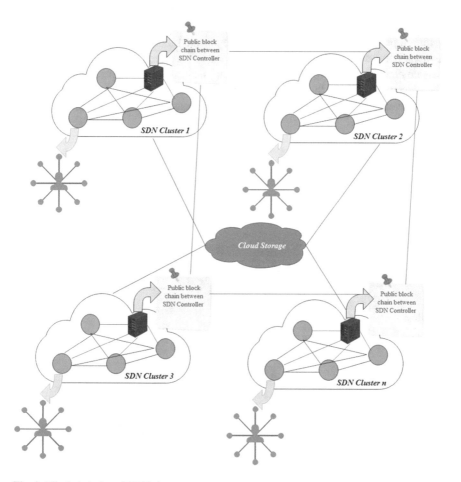

Fig. 1 Blockchain-based SDN cluster

produced, they will have full access to the history of transactions processed in the network. A distributed blockchain has no structured trade-off between public and private involvement, and several different kinds of organizations have complete access. One of the downfalls that come with a shared blockchain is the need to retain vast quantities of computing resources in order to operate a shared system. We have suggested three different approaches to solve this particular problem in a logical order. We have solved many problems with energy and computing power constraints in the DAO, and we have also eliminated the need for a POW in the form of Ethereum mining. We have also removed the computationally costly (and power-consuming) mechanisms in the DAO in order to provide improved protection. With the need for a POW mechanism, SDN controllers are first connected to the network, which significantly decreases the latency and power consumption in the network. Then, the SDN controller with IoT devices is trained into the main chain.

Even from IoT devices, I think resource constraints are not really in a way that can be extended to broad and complex anti-blacklisting initiatives or security frameworks, so we just used a private blockchain in each SDN cluster to handle the required processes. In the architecture, the private blockchain is located in an SDN domain between both the SDN controller and IoT devices, as presented in Fig. 2. A private blockchain network consists of an application and is authenticated either by the person who started the network or by a list of rules set by the person who started the network. Within each SDN region, we used the required authentication

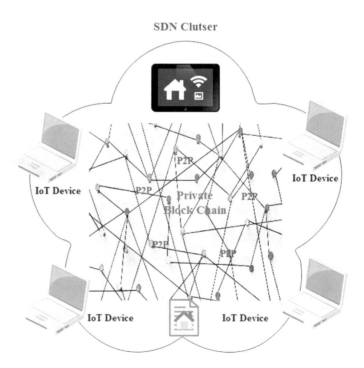

Fig. 2 Private blockchain

mechanism at each SDN controller. All private blockchains use an authentication and validation method, which is considered to be an essential form of security.

The existence of the entities participating in the transactions is transparent due to the uniqueness of private blockchains. These persons are permitted to have access to data that are important to them. There is a private blockchain in each SDN domain that retains records of transactions and has an energy-saving policy to implement input and output transaction policies for IoT devices. The SDN-Key is connected to an organization, which will authenticate all transactions that happen in the given SDN-Domain. Every CH (controller unit) has a specific public key (credential) that establishes authorizations and verifications required to generate new frames or adding new SDN regions so that other SDN controllers can authorize the merge from the block converter. If the system becomes unable to sustain adequate energy consumption or benefits, the system is freer to reconfigure its integrated SDN system. On the basis of a network controller's identity, an IoT device in the SDN domain, the cluster head stores its identification as part of a transaction on the public blockchain. The CH and the SDN controller can share a public key and a private key to carry out transactions with the computer and controller of a particular IoT system. Whenever an IoT system wishes to migrate out of a particular cluster, the new cluster has to be a part of the system's choosing climate. To ensure the protection of the IoT system, the CH checks the public blockchain for the signers of the transaction information. When the CH of the old IoT system has already stored the public key relevant to that of the IoT system, interactive chat with the CH of the next IoT system is done, and through that interactive chat, the public key relevant to that Internet system is obtained. The cluster also recognizes the new Internet of things (IoT) framework after receiving this key, and the contract is recorded in the public blockchain. In addition, each blockchain produces an infinite transaction time-based database, which can be linked to other levels to provide various kinds of assistance in the SDN realms.

3.1   Transfer of Files Between IoT Devices in SDN Cluster

Each IoT (Internet of things) device has a public key and private key inside our proposed framework so that efficient transactions could not be made from the devices of their loved ones. If a message is to be sent by an IoT device to its cluster, it's reconfirmed and then volunteered using the public key and then encrypted using the private key. Since the sender's public key is being reviewed, the authenticity of the packets of data to be sent is being tested by other participants. If an IoT computer is approved to allocate a file with members of the cluster, it will be authenticated by other members of the cluster. The request will then be stored in a private blockchain, which will be submitted by the device. If the public key has been released and the IoT device sends a folder to a receiver exterior of its cluster, the SDN cluster grasps that there is an IoT device in another cluster and enlists the cluster head to participate in the public key exchange. Finally, the file stored in the

IoT server will be forwarded to the recipient following to the IoT device has been transferred to the appropriate cluster. Alice would like things to be kept private in clusters (C1, C2, C3, and C4). In other words, she requires the private blockchain to exist even in the limits of a cluster (C1, C2, C3, and C4). After completing four stages, the remaining phases will be considered in the next phase of the file transfer.

- Phase 1: C1 needs to transmit its document to C4.
- Phase 2: The transaction is authorized by C1, the public key is released, and the transaction's private key is verified.
- Phase 3: The block is distributed over the entire system using the public key. If there is any other IoT device that does not exist in this cluster, the IoT gadget is sent to another cluster by the SDN cluster.
- Phase 4: An IoT device on the system verifies the transaction as per the public key. In reality, the sender verifies the packets by taking the public key into account.
- Phase 5: A block is attached to the string, making it transparent for transactions.
- Phase 6: The file is passed from the source to the recipient, and the file can only be decoded by the recipient.

3.2   SDN Cloud Storage

Blockchain-based SDN implementation is to provide additional safety, which is securely distributed, stable, and efficient for cloud processing applications. This statement introduces a shared, stable blockchain with an SDN-based cloud processing environment, which is shown in Fig. 3.

A. Perception Layer

At the base of the targeted system is the perception layer, and this layer contains actual data. It explicitly communicates with the system where the IoT framework is introduced. In the perception layer, nearly all the sensors and information collection devices stay related. The sensors gather data in real time at this stage and hand over the data for further analysis to the SDN protocol. This layer recognizes the gathered information and transfers it with some classification.

B. Infrastructure Layer/IoT Networks

Via specific SDN access points, IoT routing devices (e.g., switches, routers, proxy servers, and smart televisions) can send information. The information for the IoT devices can then be handled by a dynamic SDN controller, which is provided by the OpenFlow protocol. The blockchain IoT model also offers data protection that

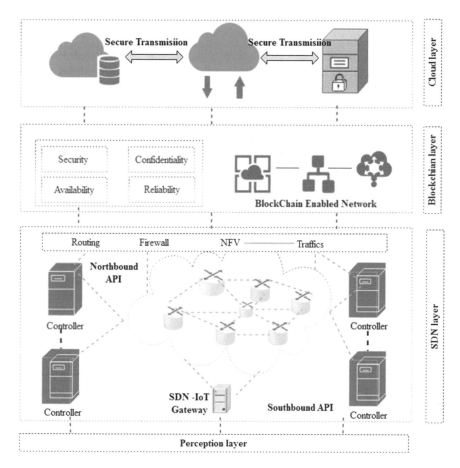

Fig. 3 Proposed BISDN architecture

allows achieving the desired layer with these data. Again, in different operations inside the SDN-based blockchain network, these data can be secure. Furthermore, in the SDN context, IoT sensor data help work effectively. In addition, the SDN environment and the SDN data plane use various distributed controllers such as OpenFlow, OpenStack, ONOS, etc., with variable sensor data so that the sensor node can be implemented appropriately in the SDN platform. Then all information is recorded privately or publicly in the cloud storage system in the blockchain networks after executing IoT data in the SDN network, with this layer acquiring extra protection.

C. SDN-Based Security on Cloud

SDN is a routing mechanism with the purpose of handling the communication program and customizing it. It can include many routing device functionalities such as technically unified control, standardized view of the network, hybrid control allocation, and network configurability. In addition, SDN provides improved network programming and control compared with conventional networking features. Compared with the current conventional networking model, SDN can also handle security risks, emerging problems, and different forms of attacks sufficiently. The SDN is also able to monitor different forms of electronic surveillance and flaws [8]. This section includes a framework for SDN that contains some great planes, such as the planes for data, control, and operation.

In the SDN setting, first of all, the data plane is the weakest plane. This aircraft allows the SDN-enabled interface to be properly linked to sensor-based equipment (router, switch, firewall, storage, etc.). In addition, it offers two different switches, such as network machines, usually running on Linux OS, connected to software-based switching devices. The other is electronic switches connected with hardware-based interfaces; inside the SDN architecture infrastructure plane, it uses the greater flow of physical device drivers. This often switches to network-based systems that are responsible for transmission, expenditure, and exchange of network packets. Moreover, the network nodes and SDN controllers are often interacted with each other by the use of more stable TLS connectors. Using the OpenFlow protocol, the data plane and the SDN controller(s) then interact with each other. After that, all protected data pass through to the control layer of the SDN architecture. Furthermore, the data plane is responsible for capturing all the information in the cloud world. The control plane (CP), then, is the major segment of the SDN architecture. It is also labeled the keystone for interaction, or the core network. A system includes elements as basic elements, such as a conceptual central and operational controller. In addition, a large interprocess communication facility is given by the logic controller. CP can also be depicted in the SDN architecture between the network and the application plane. For the desired site, it acknowledges various network infrastructures. The controller implemented several frameworks for effective contact such as eastbound, southbound, and eastbound frameworks. In addition, this controller improves the networking framework, which improves extremely customizable and reliable information in the cloud processing environment. After that, either an application layer or a plane is the top plane for SDN. Then, a considerable amount of propensity has been efficiently committed by the SDN-based system to change the forwarded routing protocols automatically. The application plane also improves Internet connections over the physical routing objects or virtual environments between the control and application plane. It then embraces network predictive analytics or advanced tasks that are supposed to be performed in massive data centers for more popular network design and maintenance levels.

D. Blockchain Networking Approach

A blockchain is a special form of database that allows various operations to be effectively implemented by a shared and atmospheric pressure system. It is depicted in Fig. 4 and completely depends on the worker's or authentication node and the node of the general participant or order.

Similarly, it can effectively provide the required device with access control and protection. Then, it serves as a stable database to bring into contracts. The central repository is often not used to maintain the activities of all people. In addition, each user uses the same storage at the recipient's end. They still hold all transaction activities and update copies at the same location to maintain the consensus mechanism. Each block can be correctly comprised of multiple transactions in the blockchain system. In addition, in each block, a hash block is clustered. Each block includes a specific timestamp, information, new hash, and past data as well as transactions that conflict appropriately. It is evident from this assumption that the blockchain technology can be incorporated in these studies to ensure network access within the proposed cloud processing architecture. Furthermore, the proposed system protection scheme would be better implemented after incorporating the blockchain method too. After that, the shared method was implemented to preserve the current blockchain-based security and multiple service access management. Furthermore, this technique is coordinated effectively with major security and enormous access management.

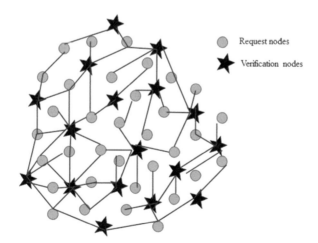

Fig. 4 Blockchain networking approach

E. Cloud Storage Management and Services

In the cloud processing world, the suggested framework improves different facilities based on a distributed blockchain method. The BC-based SDN design then offers benefits within the cloud processing framework, such as versatility, availability, protection, confidentiality extraction, and reliable storage of various properties. In particular, only blockchain cannot provide more consistency, high reliability, hierarchical controls and also improve the size for load scheduling in the framework being addressed without the intervention of the SDN method. In most cases, the connectivity of all cloud providers to all information and enjoyment services is appropriately based on network speed. First, using computation, we analyze the efficiency, durability, and resource consumption of the conceptual design. In addition, we introduce the empirical analysis of the routing protocol suggested. Then, in the suggested design, we assess the suggested routing method and also compared it with well-known protocols such as energy-aware ones. Finally, between simulation and empirical findings, we provide a comparative overview.

4   Performance Comparison

In performance comparison, we discuss an overview of the construction of the system, the testing phase, and the evaluation of whether or not the proposed framework meets the components in the test rig. The mac80211_hwsim simulator and spinoff platform ONOS have been utilized to implement the SDN cluster, and this performance is one of the pieces of proof of the research performed in the field of our SDN research. As a platform used by the Ethereum framework while developing with blockchain technology, Pythapp would be a predetermined testing framework (In Ubuntu 20.04 LTS, Ethereum platform has to be installed).

Along with related interaction in blockchain directly for the principle of the study, Pythapp was used to analyze and compare public blockchain and private blockchain. As one of the SDN domains, we have introduced blockchain and SDN for the distributed secure file or information allocating between IoT procedures. We use a virtual machine (VM) (called Ethereum) on a virtual machine (VM) (called mac80211_hwsim) with several IP addresses in other VMs, so we could easily have different IP addresses for other VMs and mac80211_hwsim to be built. An SDN-managed environment was designed and linked to the ONOS controller at start-up. The SDN controller connects to six separate devices via an open communication channel as part of the cloud network, and the devices use the ONOS controller as the CH. In every specific cluster of 15 nodes, different energy transfers are performed so that they are able to communicate through Internet-based electrical signals to transfer the required energy to other clusters. In fact, in our research, we take into account the versatility effects of IoT nodes, that is, the capacity of IoT nodes to operate in an environment that's outdoors and versatile, and also they will be a little slower. The cloud processing was somehow linked to the server of the ONOS server, which was used to handle blockchain databases, block data, and retrieval of blocks. Another study was developed to test the suggested architectural model to get an estimation of the overhead that was to be anticipated with a full implementation. We built the latest blockchain that hashes POW (proof of work) functions and processes transactions in this case (i.e., classical blockchain). However, the significant disadvantages of the IoT devices and cluster architectures are not considered by the FBC. In fact, for IoT devices, the POW (Power Over Wi-Fi) has an overhead of time and energy. Most IoT devices need to use something called an Ethernet network to bind them together. When a connection between two devices has to be created, they must negotiate a link cost before connecting any network traffic. In this simulation, we will be using the metrics throughput, packet overload, time overhead, response time, usage of resources, and use of bandwidth and delay in this evaluation. The simulation parameters shown in Table 1 can be used to conduct a simulation of the brain's reactions to alcohol.

Throughput The amount of queries for transactions between IoT devices on a network is related to the overall transaction period or total throughput time. The difference in the output of our proposed architectural solution and the FBC process can be found in Fig. 5. Since we used the cluster layout and optimized IoT node method, it enabled us to minimize overtime and storage by improving efficiency compared with other models.

Packet overload The length of packets transmitted using authentication, hash coding, and POW in classic FBC enhances the size of the payload and the packet header. While our approach uses the lower layer header and the correct IoT network radioactivity method by eliminating the POW, some evaluate packet overhead, as given in Table 2.

Overhead time This is the form of security time needed in every transaction when the requests are answered by the controller. Figure 6 shows that the BISDN method is quicker in comparison to the classic blockchain method's transfers because it

Table 1 Simulation parameters	Simulation variables	Parameters
	Simulators	mac80211_hwsim/Pythapp
	Number of SDN controllers	5
	SDN controller type	ONOS
	Number of clusters	5
	IoT devices used	95
	Simulation time	100 s
	Traffic type	CBR (constant bit rate)
	Preliminary IoT energy value	9 − 13j
	Preliminary trust value	5j
	Zone	1100 m * 1100 m
	Packet size	512 bytes

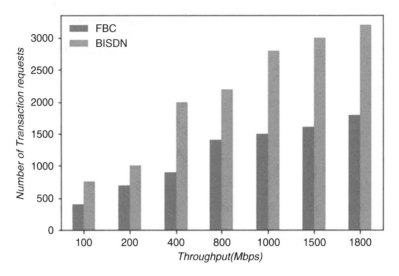

Fig. 5 Throughput comparison

Table 2 Evaluation of packet overhead

Packet flow type	FBC (in bytes)	BISDN
Controller to IoT devices	48	3
Processing in SDN cluster	56	3
IoT to IoT device	32	3
Processing of IoT device	32	3

doesn't rely on a computationally intensive encryption scheme as well as the more time-consuming POW service that this uses. In addition, the BISDN (Broadband Internet Sending System Network) approach utilizes the correct protocol for IoT nodes, which is a protocol for fast and energy-efficient data packets.

Response Time It corresponds to the transfer time of data between IoT devices because of the use of the unique routing protocol in the controller; we may notice an enhancement similar to the classic method in our proposed BISDN method. Figure 7 shows the average response time between the IoT nodes for data transmission at different sizes, where our device is quicker and more receptive than the previous methods.

Energy consumption It is the energy expended in the transaction by the controllers and the IoT system involved. The equation given below can be used to estimate energy consumption:

$$EC_{total} = Dt \times I + \left[\left(N \times Ce \right) + \left(j \times CJ \right) \right]$$

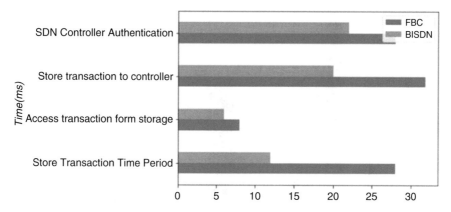

Fig. 6 Time overhead comparison

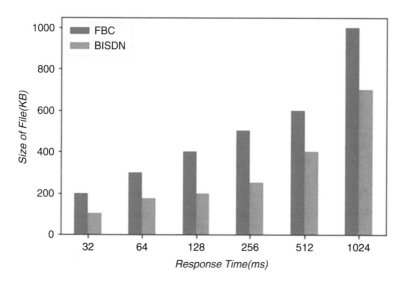

Fig. 7 Response time comparison

The FBC approach for encryption and hash produces more consumption energy by the control system and the Internet of things. In the BIDSN architecture, we have also used Data Center Networking's efficient Energy Routing Protocol for sending and receiving packets. The results of this type of analysis can be seen in Fig. 8.

Throughput and latency effectiveness Using the SDN controller, the data transmission latency in the suggested BISDN architecture can be enhanced. The effectiveness of the proposed BISDN architecture can also be evaluated in terms of delay and bandwidth, as it previously decomposes into a transfer stage of files between IoT devices in an SDN wireless domain. SDN controllers assist with key functions of IoT monitoring in each physical connection. Otherwise, in some cases, during

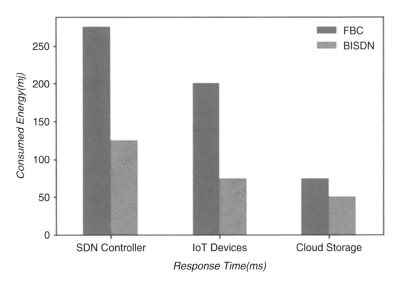

Fig. 8 Energy consumption comparison

transferring data, there is no problem when two specified IoT devices are in the same cluster.

There are also different paths in our scenario between sender and receiver that can be selected for transfer by an SDN controller. In this specific assessment, in the context of the randomized checkpoint model and constant bit rate (CBR) together, we find the mobility model and the traffic type. For this test, we also expanded the number of packets from 600 to 4500 packets with a data size of 1500 bytes to 3600 bytes. The proposed architecture performs less latency in the transmission of data in contrast to the FBC solution since the proposed architecture relies on centralization of all cluster operations. To read packets as they leave the queue and to place them in their transmission queue plus wait before sending them out, the latency of the FBC method is increased. However, the design that we envisage is more efficient since the acid cycle is clustered. In the suggested architecture, the number of different paths the data takes to reach the target shows the number of paths between the sender and the recipient. The controller of the router in the cluster often sends the data from different paths simultaneously to rapidly minimize the transmission latency. This behavior may also cause a decrease in bandwidth utilization in each cluster and ultimately trigger a decrease in bandwidth utilization during the transmission of packets. Figure 9 shows that the suggested BISDN architecture is more efficient than the FBC method, which greatly improves the performance of throughput as the SDN controller evaluates the traffic of other clusters and sends packets along the most effective route with the accumulation of packets transmitted on the path between IoT devices in the clusters.

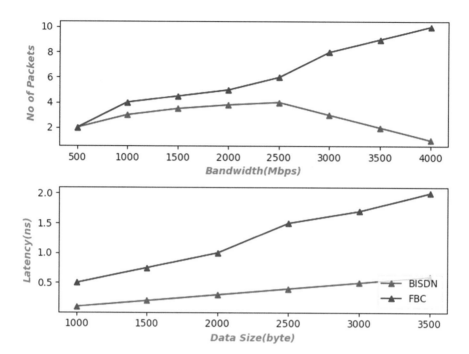

Fig. 9 Energy consumption

4.1   Assessment of the BISDN Routing Protocol

The suggested routing protocol is based on the configuration of the group and the constraints of IoT systems in terms of energy and computing. The protocol uses blockchain security mechanisms to enhance security in accordance with energy consumption standards, and SDN controls for authorization and validation methods for each group. Only, the consistency function for the suggested technique is provided through SDN controllers. The current SDN controllers are almost the same group or neighboring SDN controllers in closest neighbors, to respond to significant changes that can arise throughout connectivity in each group. As a consequence, our infrastructure offers the proposed methodology with stability. The reliability of the suggested technique is controlled by the SDN controller, as the action of the suggested technique is taken into account throughout processing in order to choose the best route to minimize energy usage.

This section discusses the findings of the analysis of the suggested routing algorithm with the existing routing protocols, namely DSDV, AOMDV, and AODV. Furthermore, the comparison of the BISDN proposed protocol for routing in Figure 10 shows the outputs of the suggested BISDN mobility situation in BISDN, DSDV, SMSN, AODV, EESCFD, and AOMDV protocols and a comparison overview of IoT protocols for cluster analysis and energy. Secure Mobile Sensor

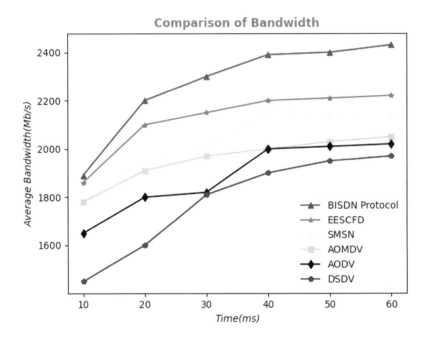

Fig. 10 Comparison of bandwidth

Network and Energy Effective Distributed Fault Diagnosis Protocol based on bounds of bandwidth, end-to-end delay, and mobility-impacted energy.

Bandwidth Bandwidth refers to the data rate that messages can be transmitted across the channel by a server. Currently, it is the amount of packages sent from the source to the target service. Figure 10 displays the bandwidth relation between the six protocols. The average bandwidth values of the BISDN protocol are higher than those of AOMDV, DSDV, AODV, SMSN, and EESCFD. They retain their value as the time is increased for every 10 s starting from 10 s and ending at 50 s, as seen in the figure. This may be partly attributed to the correct reception of packets and the smaller drop in packets. The probability of packet loss also increased with an increase of links, although the suggested BISDN protocol will help handle the links in each group through SDN controllers. The overall bandwidth is stagnation with a duration ranging from 30 s, as the sum of falling packet rises as the buffer gets complete at the time of the device list.

Average amount of delay in end to end It is the minimum amount sent from the origin to destination of the packet, which involves the pause in processing, the pause in the launch, and the period of transmission. The outcomes of this unique comparison are shown in Fig. 11.

Figure 11 shows the expected end-to-end delay against time and the latency taken from each time observed for the six routing protocols as sampling interval. It

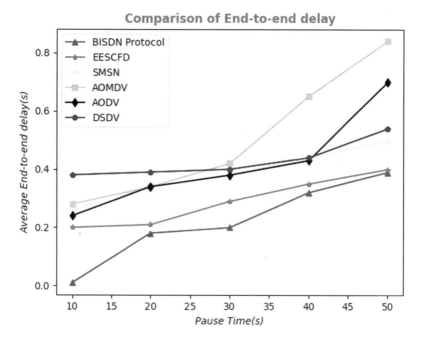

Fig. 11 Comparison of delay

demonstrates that when computation is initiated, the proposed BISDN protocol executes much less pause than AOMDV, DSDV, AODV, SMSN, and EESCFD, with a period interval from 10 s to −50 s. Due to the amount of bytes produced from each origin, the total end-to-end period rises as the sampling rate rises.

Energy The quantity of energy contained by all current IoT devices connected to the network is represented by energy. As shown in Fig. 12, the energy usage of the BISDN protocol and also other protocols is dependent on the number of links and the number of flexibility links. As the figure shows, the energy usage between both the suggested BISDN protocol and AOMDV, DSDV, AODV, SMSN, and EESCFD protocols is seen in the flexibility situation.

Higher energy than the proposed BISDN protocol is used by all five protocols. The explanation is that the BISDN protocol results from an energy-efficient method in which the relay node is modified through the SDN controller depending on the contingent node and the energy state in each group. The findings also demonstrate that the EESCFD protocol absorbs fewer resources as it utilizes a cluster-based framework in a related way. In contrast to all protocols, the DSDV protocol requires more energy.

As shown in Fig. 13, there is a weakness in the performance comparison measured using our empirical assessment and simulator outcome relation. Therefore, most fallen packets would be expected to retransmission on the same route. Overall, the findings suggest that the BISDN protocol for networking has improved

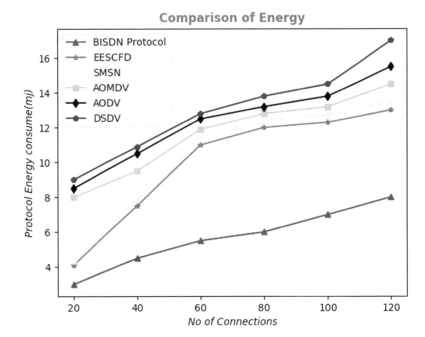

Fig. 12 Comparison of energy

performance and efficiency than the methods for group routing. This may mean that the specification for connected systems would've been suitable to be used in the BISDN architecture.

4.2 Comparison of Simulator and Empirical Results

In general, both simulator and empirical assessment are modeling methods that help provide an understanding of model success under different circumstances. In reality, the empirical model is a statistical framework that can be generalized to solve different conditions for workers, and with a particular usage case, the simulators plan outcomes and should be performed multiple times to offset the effects of computational simulations. It is expected that the simulation results will be similar to what will be found during the empirical analysis about the proposed BISDN routing protocol. Figure 13 demonstrates the performance of bandwidth, end-to-end delay, and energy indicators measured by our empirical assessment.

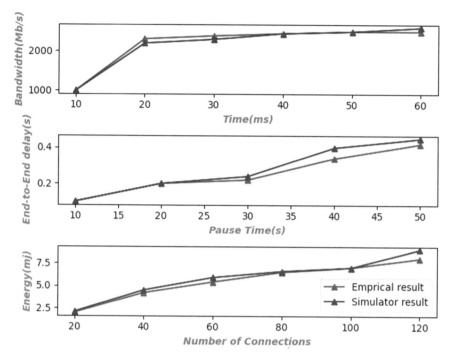

Fig. 13 Comparison of the BISDN results

5 Conclusion

There are several problems and shortcomings in terms of privacy, confidentiality, and computer capacity that would require the immediate attention of many researchers, considering the rising IoT phenomenon and the increasing demands for even more energy-efficient services. The proposed BISDN architecture for the IoT system has, two leveraging innovations, blockchain and SDN, to overcome a few of these obstacles. Customized for the IoT system, public and private blockchains are included in the group layout of the proposed BISDN architecture. By using the SDN controller's network path built for connected systems and eliminating POW from the process, our solution has helped accomplish a substantial influence on decreasing energy prices and increasing connectivity protection among IoT systems. However, there are various issues and uncertainties facing cloud computing, like stability, safety, usability, availability, access, and efficiency. In fact, certain directions and methods for the approach have been discussed by a limited group of researchers to eliminate these risks. The rehabilitation of such risks and problems at some of these times is in the growth phase of any operation and defense. The proposed BISDN framework is preferable to the FBC approach in terms of bandwidth, efficiency, and energy consumption, although it proposes an improved routing protocol that outstripped AOMDV, DSDV, AODV, SMSN, and EESCFD protocols.

In the future, the solutions of SDN provide enough features of the blockchain in the IoT ecosystem with an elevated P4 framework and evaluate its system performance with the BISDN architecture.

References

1. Abbasi AG, Khan Z (2017) VeidBlock: verifiable identity using blockchain and ledger in a software defined network. In: Proceedings of the UCC Companion, pp. 173–179, https://doi.org/10.1145/3147234.3148088
2. Aujla GS, Singh M, Bose A, Kumar N, Han G, Buyya R (2020) BlockSDN: blockchain-as-a-service for software defined networking in smart city applications. *IEEE Netw* 34(2):83–91. https://doi.org/10.1109/MNET.001.1900151
3. Alharbi T (2020) Deployment of blockchain technology in software defined networks: a survey. IEEE Access 8:9146–9156. https://doi.org/10.1109/ACCESS.2020.2964751
4. Medhane DV, Sangaiah AK, Hossain MS, Muhammad G, Wang J (July 2020) Blockchain-enabled distributed security framework for next-generation IoT: an edge cloud and software-defined network-integrated approach. IEEE Internet Things J 7(7):6143–6149. https://doi.org/10.1109/JIOT.2020.2977196
5. Khan PW, Byun YC (2021) Secure transactions management using blockchain as a service software for the internet of things. In: Kim H, Lee R (eds) Software engineering in IoT, big data, cloud and mobile computing, Studies in computational intelligence, vol 930. Springer, Cham. https://doi.org/10.1007/978-3-030-64773-5_10
6. Li W, Tan J, Wang Y (2020) A framework of blockchain-based collaborative intrusion detection in software defined networking. In: Kutyłowski M, Zhang J, Chen C (eds) Network and System Security, NSS 2020. Lecture notes in computer science, vol 12570. Springer, Cham. https://doi.org/10.1007/978-3-030-65745-1_15
7. Mohammed AH, Khaleefah RM, Hussein MK, Abdulateef IA (2020) A review software defined networking for Internet of Things. In: 2020 International congress on human-computer interaction, optimization and robotic applications (HORA), Ankara, Turkey, pp. 1–8, https://doi.org/10.1109/HORA49412.2020.9152862
8. Sharma PK, Chen M, Park JH (2018) A software defined fog node based distributed blockchain cloud architecture for IoT. IEEE Access 6:115–124. https://doi.org/10.1109/ACCESS.2017.2757955
9. Hu J, Reed M, Al-Naday M, Thomos N (2020) Blockchain-aided flow insertion and verification in software defined networks. In: 2020 Global Internet of Things Summit (GIoTS), Dublin, Ireland, pp. 1–6, https://doi.org/10.1109/GIOTS49054.2020.9119638
10. Wu J, Dong M, Ota K, Li J, Yang W (2020) Application-aware consensus management for software-defined intelligent blockchain in IoT. IEEE Netw 34(1):69–75. https://doi.org/10.1109/MNET.001.1900179
11. Rafique W, Qi L, Yaqoob I, Imran M, Rasool RU, Dou W (2020) Complementing IoT services through software defined networking and edge computing: a comprehensive survey. IEEE Commun Surv Tutorials 22(3):1761–1804. https://doi.org/10.1109/COMST.2020.2997475
12. Luo J, Chen Q, Yu FR, Tang L (June 2020) Blockchain-enabled software-defined industrial internet of things with deep reinforcement learning. IEEE Internet Things J 7(6):5466–5480. https://doi.org/10.1109/JIOT.2020.2978516
13. Rajmohan R, Ananth Kumar T, Pavithra M, Sandhya SG (2020) 11 Blockchain. In: Blockchain technology: fundamentals, applications, and case studies 177
14. Agrawal R, Chatterjee JM, Kumar A, Rathore PS (2020) Blockchain technology and the Internet of Things: challenges and applications in bitcoin and security. Apple Academic Press. https://books.google.co.in/books?id=FCoMEAAAQBAJ

15. Kalaipriya R, Devadharshini, S, Rajmohan R, Pavithra M, Ananthkumar T (2020) Certain investigations on leveraging blockchain technology for developing electronic health records. In: 2020 International conference on system, computation, automation and networking (ICSCAN), pp. 1–5. IEEE
16. Julie EG, Vedha Nayahi JJ, Jhanjhi NZ (2020) Blockchain Technology: Fundamentals, Applications, and Case Studies (1st ed.). CRC Press. https://doi.org/10.1201/9781003004998
17. Manju Bala P, Kayalvizhi J, Usharani S, Jayakumar D (2018) A decentralized file shareing & data transmission in peer to peer communication using edonkey protocol. Int J Pure Appl Math 119(14):1027–1032

Blockchain Technology Use Cases in Healthcare Management: State-of-the-Art Framework and Performance Evaluation

S. Usharani, P. Manju Bala, R. Rajmohan, T. Ananth Kumar, and M. Pavithra

1 Introduction

Ledger is a digital asset that is a shared blockchain that allows users to exchange trustworthy and validated knowledge. Blockchain can offer encryption as it increases the trust level between two parties or organizations. Blockchain proposes new solutions to old issues. By using digital currency, individuals can deal with one another without third-party brokers.

The public ledger record is owned by all members of the scheme. Due to the obvious design of the blockchain, it is not ordered to manipulate the actual ledger. Via crypto-algorithms, we can accomplish this goal. A distributed ledger data model is a set of ciphertexts that are in a rigid, consistent, and constant sequence. A cryptosystem is used for integrity. Any transaction references a preceding transaction in the ledger and has a rigid sequence to what will be documented next. For example, the blockchain network is the first stage of the chain. This is the first chain released by Bitcoin. This style of arrangement makes it easy to trace any member of every interaction. The systematic control empowers law enforcement to discover and mitigate illegal narcotics peddling practices. There are various kinds of blockchain systems. These are considered open and commercial blockchain-based. Open blockchains like Bitcoin render several improvements to existing technologies. Restricted blockchains (such as Hyperledger fabric) restrict the quantity of permissioned members to execute an agreement. A reputable medical firm has to be listed on personal blockchains to check the consistency and efficacy of their goods. An impression of a blockchain organization is depicted in Fig. 1.

S. Usharani (✉) · P. Manju Bala · R. Rajmohan · T. A. Kumar · M. Pavithra
Department of Computer Science and Engineering, IFET College of Engineering,
Villupuram, Tamilnadu, India

© Springer Nature Switzerland AG 2022
P. Raj et al. (eds.), *Blockchain, Artificial Intelligence, and the Internet of Things*,
EAI/Springer Innovations in Communication and Computing,
https://doi.org/10.1007/978-3-030-77637-4_7

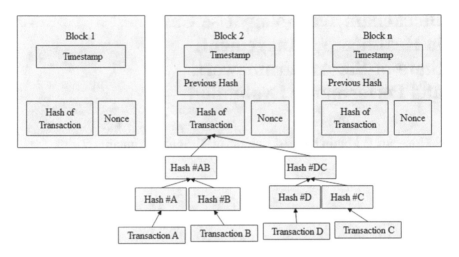

Fig. 1 Structure of general blockchain

2 Related Work

EHRs are traditionally used by health facilities using a client–server architecture whereby sole governance is maintained by the hospitals [1]. Various records disperse the diagnostic details of an individual seeking care from various hospitals. There are many fog health apps proposed to solve this problem. In client–server and fog architectures, nevertheless, confidentiality and protection are the primary issues [2–4]. Through handling the medical data on the cloud and storing the hashing of the data in a unique blockchain network, many research initiatives resolve the privacy issue [5–8]. Many reasearches have done in theoretical frameworks for data confidentiality and integrity in a blockchain of health information. Several reports suggest a clinical data processing scheme focused on blockchain, including several clinics. A blockchain-based interoperable technology helps related health practitioners to share the health records of patients. An Ethereum-based smart contract blockchain platform provides access to medical data [9–11]. However, the energy-hungry and quasi-Proof of Work trust method is used for public blockchains. An EHR blockchain-based architecture uses an agreement on Realistic Mediaeval High Availability that is more environmentally friendly and looks nicer than PoW [12]. Health data management systems enable patients to preserve general healthcare data management, and only physicians are entitled to access their data over the network [13]. In these schemes, only the person's health history can be obtained by associated health practitioners. A PHR blockchain healthcare scheme and information update control for all primary care providers and patients. As they know, there seems to be no research that contrasts data processing frameworks for client–server and PHR blockchain healthcare. In this chapter, a PHR blockchain healthcare platform is introduced, and its execution time and volume of data exchanged are contrasted with the client–server device paradigm for updating and querying health information. This is for a growing network of hospitals and documents.

3   Challenges in Healthcare

There are numerous problems that face HAIS in Singapore and other Asian countries.

3.1   Knowledge Confidentiality

In terms of experimental research in medicinal experiments, drug studies are set up to test medication's efficacy. Details on numbers, test outcomes, progress tracking, and so on would be registered throughout the medical study, and they can be updated if appropriate. This will affect the credibility of a company inside federal enforcement authorities or customers.

3.2   Prescription Referential Integrity

In drug companies, medication piracy is the most significant issue. The main issue with illegal narcotics is that they do not have the necessary components as needed and do not work. This won't really heal the illness but can be used to relieve symptoms.

3.3   Patient Illness Interpolation

These public hospitals include many illnesses and medications for clinicians. Appropriate paperwork is important for adequate drug care. A further problem with this project is that it is cost-intensive and challenging. To conclude, they are busier and less stable.

3.4   Big Data

There is an enormous amount of information being produced and distributed through various stakeholders. As it is conveniently feasible to access the evidence and create improvements, this makes the platform a scam. It seems that there is no "arbiter of facts," which the organization will use to enhance customer care.

4   Blockchain Predictive Model for Healthcare Management

Blockchain is a kind of shared ledger that has helped build and decentralize the economy, as a result of which a number of commercial operations have been aided considerably.

When a connection occurs between the participants, the system processes payments. However, there will not be any redundant records in the blockchain after a block is registered. The blockchain database is not maintained on a dedicated processor as is usual in current cloud servers but is spread among all the machines of the system, named checkpoints. The overwhelming significance of this innovation lies in the devolution of agreement in the database. The devolution of agreements offers the cost of publishing that can raise the risk threshold of any form of financial operation. With the second interpretation, a network such as NXT LIVES will be extremely secure and more stable because of the verifiable replication of identical blockchains across the nodes of the network. Because of blockchain design, this technology is not protected by a centralized framework. Important research confirms this argument by demonstrating the transferability of innovation in various domains like banking, economy, finance, insurance, trade, and agro-food to the benefit of various disciplines.

With the rising trend of the use of consortium blockchain, this platform will become a significant tool to fight syndromes. The unexpected emergence and the swift and unchecked dissemination across the globe of COVID-19 have pictured the inability of the present medical protection infrastructure to contain community healthcare crises, but also an apparent absence of sophisticated forecasting mechanisms focused on the exchange of scientific evidence at a fast pace, capable of mitigating such crises. Blockchain can be applied within the healthcare industry primarily for exchanging and precise awareness of people's information, telemedicine data and, in some cases, inventory governance of pharmaceutical products and medications, the governance of pharmaceutical supplies, enhancing biomedical science and intellectual property, and the progression of clinical discovery.

Improvements in the practice of surgery result in ever safer surgeries. The productive use of innovation can include the collection and sharing of patient data between hospital systems. There is a chance that the security of health records would be strengthened by using ledger technologies. The problem of integrations between various electronic healthcare records (EHR) can be resolved by enabling integration in the center of multiple distinct ledger systems. "Blockchain" is also used to describe a "distributed ledger" method. Information engineering could positively foster the growth of accurate healthcare.

They are utilized to streamline and automate risk assessments, enhance production lines of generic drugs, and check their safety and comply with existing legislation. Among the things that obstruct technological progress is the lack of proper IT facilities to exchange empirical testing findings and drug trials. Blockchain is used in medical research for information storage, distribution, and consumption.

This distributed and open software leads to the questioning of privacy protection for patients and user privacy against police, especially about the secret exchange of confidential information in distributed ledgers.

It is about involving biomedical knowledge and health-related knowledge via the use of ledger technologies. In this situation, the medical professionals and organizations create an innovation model. This would be achieved via a review of the medical diagnosis outcomes and the use of the evidence to forecast possible medical consequences. This kind of tool can add to the broader phase of healthcare management to incorporate the medical result of stress, depression, and anxiety.

Blockchain technology of artificial intelligence can create valuable data available in clinical treatment (risk assessment). Last year, scientists implemented human cognition and data science to aid them in determining accurate diagnoses and medication of many pathogens. The services offered by blockchain technology would not disclose any private details of clinicians and recorded medical records. These clients may take part in information analysis systems to build statistical frameworks on clinical management or epidemics emergence and growth. Frameworks that are assisted by artificial intelligence can be changed into an integrated context of knowledge interacted in the framework. This template will be revised and improved to be the most accurate. The image recognition framework ceases, and the final agreement template is found (Fig. 2).

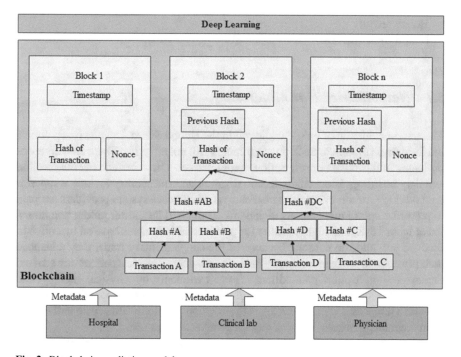

Fig. 2 Blockchain predictive model

The need for ledger and machine intelligence-powered models to establish epidemics surveillance is profoundly useful and might relate significantly to the prevention of epidemics emergence and spread. The powerful effects of diagnostic and medical knowledge relating to clinicians, if continuously modified, can have a considerable impact on service confinement strategy at local and global scales.

5 Distributed Ledger for Public HealthCare

The blockchain technology use cases have been identified each day. With them, the whole health service can be completely remodeled. Numerous healthcare and distributed ledger companies have published blockchain-based methods to increase health services to both healthcare providers and patients. Blockchain is revitalizing the field of healthcare due to the decentralization of patient health records, distribution of and monitoring pharmaceuticals, and easy payment options.

- Narcotic medicine monitoring
- Data distribution with tele-health to improve outmoded attention in developing countries
- Cancer controlling ideas
- Cancer registration sharing
- Digital patient uniqueness
- Health records
- Health right decision

5.1 Narcotic Medicine Monitoring

There is an epidemic of opioid addiction in the United States. Despite being done a lot, the current system to track prescriptions is still not effective. There are numerous problems in the existing prescription opioid market. These include overmedication and data hoarding. Blockchain decentralization and traceability include a superior methodology for the tracking of medications. Healthcare system providers are paid to prescribe opioid medications to patients. There will be shorter production hours, lower labor costs, and higher profits for providers. Pharmacies have an incentive to produce and allocate narcotic because of the supplementary trade, they advanced their profits, and make more money back to the owners. Besides, patients are advised to devour prescription pain killers. These treatments by physical therapy and postsurgery recovery are often disappointing and frustrating to the patient. Opiates provide a temporary relief, but it leads to addiction over time. This cycle will benefit from technological solutions that align incentives.

A blockchain-based infrastructure will set up a secure connection of hospitals and clinics to manage drug-related purchases in an open and transparent position in order to counter such opportunities to lead to the growth of the opioid crisis. Without the confidence responsibility among each other, this will enable poorly coupled physicians to enter other information silos. They are also urged to incorporate newcomers into the scheme, so they can build a much more system consisting of each current member. Laws should be unanimously coded to allow the partnership to start incorporating new providers. This type of environment will remedy some of the current problems in the heroin system by not relating to the specific content of a drug transaction. To track over medication of narcotics to offenders and to discover habits of writing prescriptions, a detailed prescription record will be made accessible. Instead of access to evidence that will establish important analysis, services would have to fit the threshold to enter the partnership. Patients can access medication that is acceptable to the situation by monitoring the record of opioid overdose, and therefore patients can be deviated from either the risks of narcotics toward more successful treatment choices.

5.2   Data Distribution with Telehealth to Improve Outmoded Attention in Developing Countries

Telemedicine is expected to have a major impact on remote areas because of how easily accessible it can be and the simplicity of it. Patients nowadays prefer to undergo treatment in facilities that offer convenient or time-saving medical services. This lowers patients receiving emergency care upon referral of mild, however, quality requirement. Mobile and telemedicine devices have increased the accessibility to healthcare and these companies offer 24 hour, year-round accessibility to medical care and their user-friendly health applications. For instance, health applications permit patients to check their blood pressures and store them in their iPhones. If required, this information will then be recorded to the provider.

It is easier to access healthcare when it is done through telehealth. Using such on-demand providers reduces the continuity of patient care as patients may be treated by different professionals. A real factor is that even if clinical information obtained during occasions of telehealth treatment is unavailable by healthcare providers for patients, it generates unreliable health information and threatens the quality of healthcare services.

The potential of blockchain technology in bridging the barriers between those technology benefactors of blockchain technology unaided cannot report all the information of privacy challenges as it needs to be integrated with government and healthcare records. Figure 3 is a high-level conceptual infrastructure showing how these disparate systems and data including blockchain can be connected (represented as cylinder database objects).

The "open architecture" of databases opens up a new channel for the access of data from other systems. In order to have a permanent record of all transactions, a smart contract that will control transaction data in between networks based on contractual arrangements is often used. In reality, far beyond seen in Fig. 3 would need to be included.

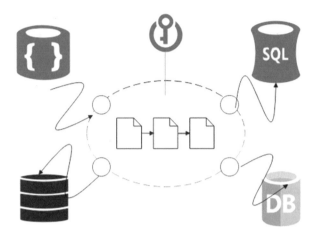

Fig. 3 A great level of architecture established on the blockchain to link diverse health data warehouses and record the background of data sharing

5.3   Cancer Controlling Ideas

So treatment plans for cancer patients are rarely either taken or rejected. Getting fresh views of specialists might help you decide on the course of action for this cancer patient. In India, several hospitals have had at minimum one cancer panel to evaluate and address the cases and care facilities of particular cancer patients. Moreover, family members and cancer patients seldom take part in the decision process. Narrower medical facilities would need professionals than even a bigger facility from a wider variety of backgrounds. Owing to the disassociation of people, enhancing the standard of living for people with cancer could be overlooked.

In veracity, healthcare consumers might request to change providers for a second opinion. Without fully accessing key health information about oneself, it is impossible to participate fully in healthcare. Now all the documents will be sent directly to the service supplier. Patients with serious disorders will be less interested in data sharing systems in this highly sophisticated community to get information circulated as rapidly as feasible to reduce service complications. A function of patient-controlled exchanging data that is absent from current healthcare organizations would enable people with cancer to discuss multiple alternatives just before the first chemo treatment is received.

Blockchain technology provides the ability for permissioned sharing and decentralization rather than establishing a new trustworthy third guy, enabling established mutual trust to be collated and transmitted through several organizations and suppliers. That's also suitable for the new prescription mechanism for hospitals, in that it will be given access to the system of each provider. Throughout nations, states, and regions, they could be controlled. Blockchain would effectively allow confidence exchanges among providers and patients to be recorded. Patients will have the right to decide which provider can have access to their medical records.

5.4 Cancer Registration Sharing

Data exchange is important in cancer care since these therapies are typically nuanced and have cure rates that are only yet another. Being willing to exchange cancer knowledge actually protects the credibility of clinical research outcomes to guarantee that it is not a product of random chance. Dispersed medical inquiries work to attain a substantial population dimension, thus facilitating cancer research and eliminating time delays. In the United States, only about 3% of cancer patients are being studied. Throughout this way, most cancer patients are treated by observing on a limited population of patients with a related census, family records, supplementary detection, and so on. Cancer records office make very basic data from the occurrence of cancer in different parts of the country or specific geographic areas. Cancer registries are currently decentralized and unable to exchange information because of a lack of interoperability, which can be solved by blockchain technology. Information from many patients can be used by artificial intelligence to build predictive models, which can assist clinicians with their decision-making. Blockchains can be used to design an event to share analytical prototypes and enhance the precision of studied health ideas.

5.5 Digital Patient Uniqueness

Patient identification matching is another essential component in the health information exchange. Institutions such as the Health Records Registry and the Enterprise Health Records Registry have been developed within a healthcare system or among trustworthy networks to handle patient personalities. In spite of the increase in improvement investment, correctly matching patient's data residues is challenging. Mismatching of patient identities leads to mirrored health history and missing or false documents. According to a report by the Office of Human Research Safeguards of the Centers for Medicare and Medicaid Resources of the United States, there are around 195,000 and 320,000 avoidable diseases in United States per year medication errors. In addition, healthcare organizations that hold these overlapping documents to address wrongly combined mistakes and patients that undergo continued support or medication problems are subject to substantial expenses. These mistakes often impair compensation when "obsolete or inaccurate data" requests can be rejected to never consider the protection danger, so patients reveal their private information. Some care institutions use different systems to collect personal information, even for the same patient. The demographics information, such as names, dates of birth, contact information, and the SSN, will help identify a patient. References can be stored in various ways, such as legitimate first or last identities, nicknames, and then last identifiers, with or without a second name, and similar or identical names can be exchanged by patients. In addition, patient data entered into the database automatically will involve spelling errors or mistakes, so the more medical care entered, the greater the chances of mistakes. While the identity by each patient can be broken down into a single specific type, it normally may not transfer through institutions. But without the framework of patient recognition functioning or integrated, it is very likely for identities to be incompatible across care sites. The nature of the blockchain network incorporated a dispersed individuality system. Numerous current blockchains practice cryptographic discourses to signify individualities. That identity is statistically connected with a secret identifier that can be used without sharing any personal data to easily process and check the authenticity of an identity or address. Its decentralized and transparent characteristics can use and apply consistent secure individualities for patients across all health vendors within a nation and beyond. In case patients change their addresses, their new ZIP codes are also simple to list and change.

5.6   Health Records

Unlike the typical electronic health records (EHRs) that hospitals use today, PHRs are designed for patients to assist them in managing their own health records. The main aim for PHRs is to help individuals gather and handle the decent health data gathered from either a number of sources easily and securely. Through enhanced patient care information management, patients and family members can monitor the utilization and exchange of health information, validate the integrity of their health data, and correct unintended errors. Other businesses have begun to use large enterprise systems such as Health apps and HealthVault app. The privacy dilemma and the absence of interoperability problems at its heart are not addressed by unified approaches. Thus, clustered methods can face problems comparable to current decentralized EHR schemes.

By contrast, blockchains give decentralization and dispersal of the device of financing to persons through such mechanisms. An extensively accessible and safe fact dispersal service will connect hospitals and patients without copying every healthcare provider they visit. People could very well connect their personal smartphones to medical data on blockchains. Digital contracts should be created to ensure that patients are in charge of their access to data, are mindful of the origins of consolidated datasets, and are notified when services are processing their data. Via investigations logs, patients are informed of the sources of the data and access background.

5.7   Health Right Decision

To cover people and their belongings, health insurance is used. With health coverage, expenses such as doctor visits, medical, and surgical will be provided by the insurer. Patients may pay a portion of their medical bills upfront, but the remainder of the bill is paid out of pocket because insurance is in place. The provider's claim is then approved and reimbursed by the insurance company as an insurance payment.

The claimant may decide to completely pay the statement, announce that the assertion is not approved, or minimize the amount charged to the covered. For example, in the healthcare system, there are procedures that are not likely to receive funding because they are unlikely to provide any benefit to an individual. Ensuring correct encoding of all statements is critical. When a health insurer accepts a petition, a comprehensive analysis begins. Minor mistakes in patient records, such as spelling, can contribute to the denial of the claim. Due to today's large volume of claims, adjudication is difficult. Almost 22% of claims tend to be rejected because of incomplete or incorrect data or missing proof of services claimed for. Smart contracts have the potential to automate adjudication further by translating disputes into executable programs. There are many benefits of creating preestablished agreements via smart contracts.

6 COVID-19 Healthcare Use Cases in Blockchain

Due to the potential of emerging technology to enable enhanced data reliability and verification, the core benefits presented by blockchain are associated with increased trust and security. At the simplest stage, blockchain transfers ownership and information control from one unified source to multiple information sources. Three examples of blockchain use related to COVID-19 in healthcare are as follows.

6.1 Reference Monitoring

In order to track the potential spread of the latest coronavirus, many countries have concentrated on guideline surveillance wherein infected individuals are allowed to list all the other persons they may have interacted with for a specified period. Decentralization of data tends to facilitate critical healthcare practices, such as benchmark testing, as the system focuses on the use of concentrated, sensitive data to alert health agencies of who is participating, depending on their activities and connections, and who may be at risk of exposure holding that the privacy of people is important in reference monitoring.

6.2 Medical Record Exchanging

Another important use for the ledger is the gathering of patient data information after an illness or crisis and set up a "warm" electronic health record as it relates to COVID-19, which may be used by multiple types of organizations to share patient details while handling unidentified patients during a contagion or other crises and natural catastrophes. A rather forum would allow physicians to connect with patients being unable to access their usual physician, so they are also providing the full range of necessary treatment and prescriptions. The central concept of the technique is that patients' electronic health records follow them wherever they choose. In several other terms, their private health information is also a direct challenge of where the patients live after a tragedy because they are able to obtain the medical services required. These medical data can be supplied by a ledger virtual wallet, ensuring entry, anonymity, and communication authenticity.

6.3   Benefactor Accreditation

Benefactor accreditation, which is the process of reviewing clinicians' credentials, training, and knowledge, is an often routine, moment method and can lead to care disparities that resulted in insufficient clinical outcomes for both practitioners and insurers. By using the blockchain for the method, benefactors will maintain control of their own data, which allow the exposure of healthcare networks, insurers, and other organizations to their data as they like. Earlier last year, five organizations announced intentions to use the new blockchain work under the supervision of ProCredEx to reduce risks and costs involved with the traditional accreditation process through the use of blockchain technologies.

7   Blockchain: Ethical Concerns

The use of blockchain and other technologies is correlated with a variety of ethical problems.

7.1   Smart Agreements

Blockchain is one of the evolving uses for smart agreements as a network. These are self-executing digital agreements, consisting entirely or partly of software code, which are implemented on a shared database and can be used to perform transactions on the database between individuals. In theory, smart agreements are subject to the normal principles of commercial agreements, and these types of agreements give rise to legitimately binding responsibilities if the conventional procedures for concluding agreements are met (i.e., suggestion, reception, goal to produce authorized relations and deliberation). In England, by the Electronic Commerce Act 2000, the use of digital signature to conduct agreements is legal (the 2000 Act). Personal keys used on blockchain systems that are meant to verify records or payments are expected to fall under the "digital signatures" concept found in the Act of 2000.

7.2   The Protection of Data

To the degree that facts maintained on a blockchain database might comprise private information, a dispute may arise between blockchain technology and data privacy regulations, in particular with respect to enforcement with the General Data Protection Regulation (GDPR). In turn, this would be compared with the actual usage case at hand, and each scenario will also have to be judged on the basis of all its unique data.

8   Blockchain Healthcare Benefits

The most common blockchain technology in healthcare applications at the time is holding our critical clinical data secured and protected, which is not unexpected. In the healthcare sector, privacy is a big concern between 2010 and 2020; over 226 million data attacks were reported in medical records. Credit card and banking data as well as medical and genetic research documents were stolen by the offenders. The following advantages compared with the client/server method are offered by a blockchain-based medical platform.

- Data integrity: It is one of the key blockchain features. It is never possible to erase something that is documented in the database. This enables all the details to remain preserved for medical applications.
- Confidentiality: The blockchain network enables the implementation of default access control rules, which, in conjunction with various activities, grant and (or) restrict this access.
- Faith: Blockchain's decentralized design and agreement algorithms establish trust within a blockchain platform; based on these algorithms, choices are made to allow or give access to information on the network.
- Responsibility lenience: In a client–server system, a central record handles the health data of patients. It is not possible to recover them until the data are destroyed. In fault tolerance, the replication characteristic of blockchain aids.
- Data sharing: In the new client–server structures, the information of a patient is distributed through the databases of many hospitals. A dynamic process is the exchange of data between various clinics and medical organizations. However, the patient information registered in the database is shared across all the hospitals in the system on a blockchain-based framework.
- Integration: In a client/server-based environment, each patient uses interconnected data models and frameworks to store patient data in a different repository,

resulting in compatibility issues. This dilemma is overcome by the synchronized and mirrored database in the blockchain.

- Exclusion of duplication of tests: Medical data are widely dispersed among multiple healthcare employees. A patient also has to undergo recurrence-specific testbed and clinical exams. Not only does this cause high medical costs, but it has opposing possessions on the mortal body as well. To escape medical examinations, the mirrored blockchain database helps.
- Higher security: Numerous cyber threats such as malware and intrusion are vulnerable to the current client/server-based infrastructure. By producing a fraudulent identification or mixing a patient quantity with a fraudulent benefactor to privilege health coverage, stolen health records may be used to purchase medical devices.

9   System Framework

In a wider sense, healthcare is dealt with the PHR model. The present research, on the other hand, extends and makes updates to the blockchain-based design and integration of PHR, as well as tests the PHR design in three other situations of production health organization. Healthcare based on the blockchain infrastructure of PHR and the effects of healthcare data duplication is explained in detail below. The following two structural levels are composed of the PHR blockchain architecture framework [4]:

(a) Client systems built in healthcare providers and medical facilities
(b) Server layer deployed to a blockchain-based platform

This framework is generated by a personal peer-to-peer network, which organizes medical records into data blocks consisting of a relational database and a shared health database. Figure 4 portrays the PHR model structure. This figure demonstrates how customers interact through pull and push communicating with the corresponding blockchain technology. This format allows all clients linked to the network to constructively modify their information; that is, data blocks can be simultaneously sent and retrieved.

The blockchain network is built on the server on a group of decentralized super-peers. Inside a Knowledgebase, which is a nonrelational NoSQL repository built on a Graph or RDF DBMS, this secure network maintains databases.

In an ER (Entity-Relationship) model, the PHR design often uses a parallel repository to maintain the data blocks in analogy format, which is a relational database management system. These themes meet the pattern of EHR health information, which we implement in our blockchain platform for interaction and data processing. Archetypes' structures are the components that form the structure of the EHR health record. In constructing the PHR smart agreement, the chained health data blocks in this repository are used.

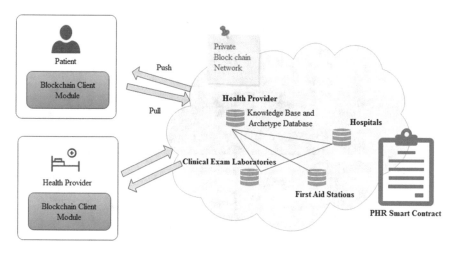

Fig. 4 Architecture of PHR prototype

Figure 5 illustrates how healthcare information blocks are chained up by the PHR model. Each data block comprises of (a) the data generated by the prototype that contains the medical record, (b) the domain comprising the cryptographic hash that represents the digital certificate of the prototype information, and (c) the hash value reference that describes the previous data block. Because it is the only node in the related list, the first data block is called the evolution block and the "preceding hash" area refers to no other data block. To validate and avoid breaches of PHR details, the PHR model introduces the blockchain smart agreement functionality. The position of each entity in the blockchain framework of medical documents is another focus of the PHR model. In general, our model only enables superpowers placed in the private field to determine the accuracy of data blocks. Consequently, participating nodes only access super peer-supplied web applications. In addition, customers often generate data that are analyzed and shared by super-peers upon this blockchain.

Data blocks can be processed in PHR models in two different ways:

(a) Distributed across all instances, adopting the method followed by the blockchain.
(b) Use a duplication mechanism, such as Note, to duplicate documents in the individual blockchain framework on those devices exclusively.

To endorse all types of duplication, the PHR design can be customized so that we could set up how much entities we want to duplicate the blocks of data while using the Chord method. To allow this choice versatile, the Chord method has been used. This versatility is one of the key features of the system, considering that replicating health frames for all entities in the system cannot be optimal or even performance art.

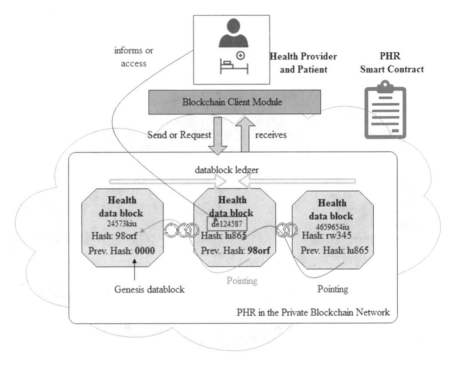

Fig. 5 PHR prototype chains health data blocks

10 Experimental Environment

The network architectures test the client/server method and the PHR blockchain design framework to recognize a chunk comprising of one health record within the PHR blockchain architecture. In a case where many records are clustered into a block that is constrained by the network capacity, this is to reduce the lag in accessing medical records. In the tests, we used a medical record of 26.85 MB in size. This is because the record contains standard-size photographs and text, such as intraoperative photographs (1.74 MB), skeleton telephoto (0.95 MB), cardiac cephalogram (1.50 MB), and photographs of body conditions (23.17 MB). A key length of 1.5MB, describes a medical records request in the framework. The heading and body consist of a chain in the blockchain. Experiments show that the body contains a medical record and the heading comprises configuration details such as the hash of the prior block, time and date, Merkle origin hash function, block numbers, and version. We use the regular block step length of 80 bytes in our experiments. SHA-512 that produces a specific 512-bit output signal input is the cryptographic hashing method included in the experiments. The choice of SHA-512 is due to its acceptance among applications of the blockchain. Both tests for the client/server and PHR blockchain technology are conducted to test the effect of a complex medical data model, with a growing number of medical records (3000,

4500, 5000, 6500, 7000, and 10,000) and growing health centers (25, 45, 75, and 100). Although holding the number of hospitals statically at 20, well maximize the amount of records, and maximize the amount of hospitals, eventhough holding the amount of records persistent at 5000. The minimal amount of data chosen is 5000, the total number of medical information staying at 10 hospitals every day which based on a study from the Center for Health Indicators. The simulations are being developed using Python Predictive Analysis Simulator.

11 Performance Evaluation

Figure 6 illustrates the computational cost for client/server upgrading medical records and the minimum PHR blockchain framework built with a growing amount of medical records. This demonstrates that with growing medical records, memory usage for client/server and PHR blockchain frameworks increases sequentially. For the client/server design, furthermore, the time complexity is less than that of the PHR blockchain. It is because of the method of agreement used for system testing and duplication in the PHR blockchain. The medical report that must be changed to the database is sent for confirmation to all hospitals with a copy of the database. Furthermore, before adding it to the database, each node will relay the chunk's hashing to some other peers in the system for agreement. In the client–server

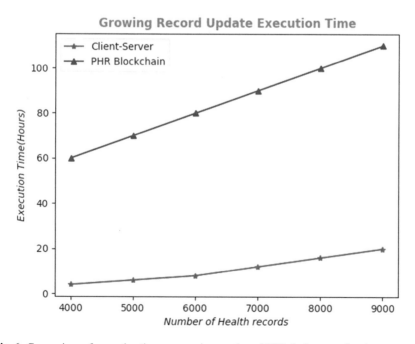

Fig. 6 Comparison of execution time vs. growing number of PHR during record update

method, on the other hand, the data validation query is sent to the hospital where the medical care information resides.

As a result, the client/server implementation period is less than the PHR blockchain technique for updating medical records. For instance, relative to the PHR blockchain network, the client–server solution proceeds 8.6 times fewer times to update medical records. In spite of the volume of data transmitted to modify medical records versus growing numbers, Fig. 7 displays the efficiency of client–server and PHR blockchain systems. This indicates that the volume of data transmitted to the PHR blockchain framework is more comparable to the system of clients/servers. This is because the demand for medical record updates is transmitted by the broadcast message to all peer nodes in the system, creating further data transmission. In particular, each peer periodically transmits the hash of the block to all other node peers maximizing the transfer of data. On aggregate, similar to the client/server method, the PHR blockchain framework shares 15 times more details. Figure 8 illustrates the client–server time complexity and the built minimum PHR blockchain-based medical models for verifying medical records with a growing number of medical records from the repositories.

This illustrates that the PHR blockchain query execution is considerably less than that of the client/server method. That's because the data are obtained from the registry on the client/server where the database resides, while the data are obtained from the global version of the database in the PHR blockchain. The time complexity in the PHR blockchain is only due to the propagation to all nodes in the system of data processing requests and the inclusion of the application as a transfer in a block

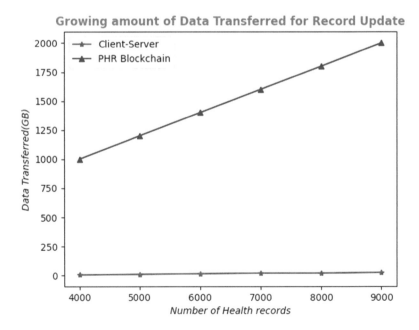

Fig. 7 Comparison of data transferred (GB) vs. growing number of PHR during record update

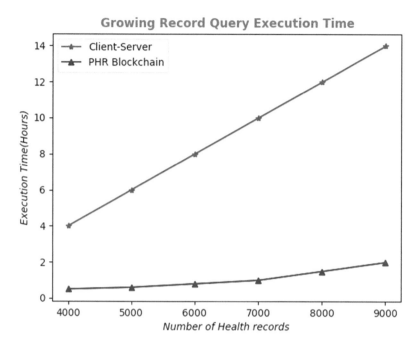

Fig. 8 Comparison of execution time vs. growing number of PHR during record querying

by agreement. For instance, relative to the client–server system for parsing well-being information, the blockchain for PHR is 11.78 times quicker.

Figure 9 indicates the volume of data transmitted from the repositories with a growing number of medical records by client–server and PHR blockchain frameworks for parsing medical records. This illustrates that the amount of information exchanged by the PHR blockchain network is more comparable to the client–server method. This is attributable to the smart agreement used by the PHR blockchain for the PBFT. In contrast to the client/server strategy for parsing health information, PHR blockchain shares 1.2 times more data on the scale. Even though, the sum of data transmitted for database request by PHR blockchain. Figure 9 is similar to the one for database upgrades as in Fig. 7. That's because a frame includes a specific health file to be changed for server adjustment, while a query request that is comparatively small in size is used for database queries.

The system performance of client–server and PHR blockchain frameworks for medical record upgrading with a growing number of hospitals is shown in Fig. 10. With a growing abundance of medical records, the relative output is close to that (Fig. 6). It demonstrates that there is more processing time for the blockchain system than for the client/server. For instance, relative to the PHR blockchain, the client/server solution takes 15% less period to update medical records. Figure 11 demonstrates the amount of information transmitted to a growing array of hospitals through client/server and PHR blockchain methods to upgrading medical records. Due to the smart agreement used by the former, it indicates that the PHR blockchain

S. Usharani et al.

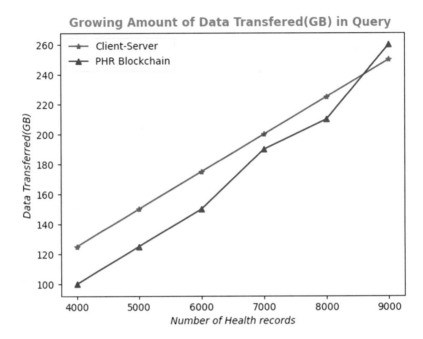

Fig. 9 Comparison of data transferred (GB) vs. growing number of PHR during record querying

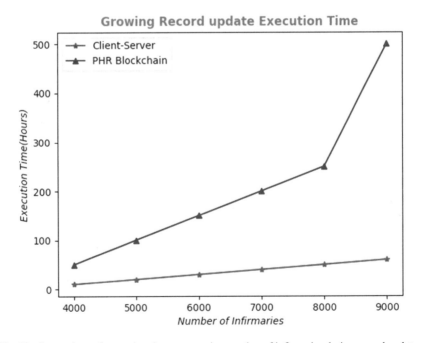

Fig. 10 Comparison of execution time vs. growing number of infirmaries during record update

execution time is more than the method of the client–server. For instance, with 10

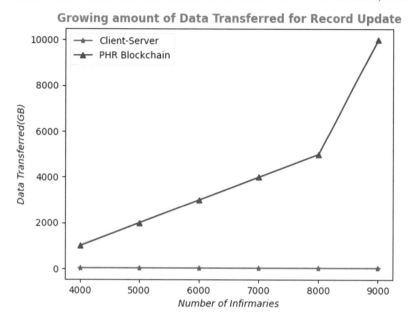

Fig. 11 Comparison of data transferred (GB) vs. growing number of infirmaries during record update

hospitals, the quantities of information exchanged by the PHR blockchain are 15 times greater than those by the client–server process.

The time complexity for parsing medical records from repositories with rising hospital numbers is shown in Fig. 12. When we raise the number of medical records, the overall production will be the same (Fig. 8). It demonstrates that the PHR blockchain time complexity is considerably less than that of the client/server method. For instance, the PHR blockchain is 10 times faster than the medical record query client/server method. Figure 13 indicates the amount of information transmitted from repositories to a growing array of hospitals through client/server and PHR blockchain systems to query medical records. It indicates that the volume of information that the client/server system transfers is static. This is because, regardless of the number of clinics, the number of medical records questioned is static. The question will be submitted to the hospital that has the necessary record. With the number of hospitals, however, the volume of data transmitted by the PHR blockchain network for parsing medical records rises. This is because of the transmission of knowledge due to the PBFT agreement between the hospitals. For instance, compared with the client–server model for 15 hospitals, the amount of information exchanged by the PHR blockchain is 1.5 times greater for 15 hospitals.

Table 1 indicates the efficiency of client/server and PHR blockchain methods with growing numbers of medical records and hospitals in spite of time complexity

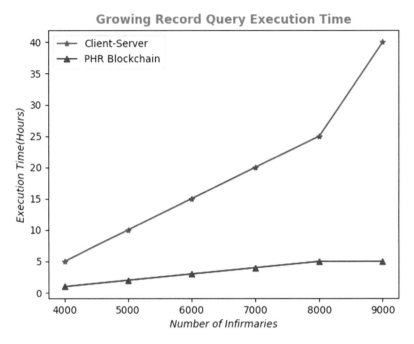

Fig. 12 Comparison of execution time vs. increasing number of infirmaries during record querying

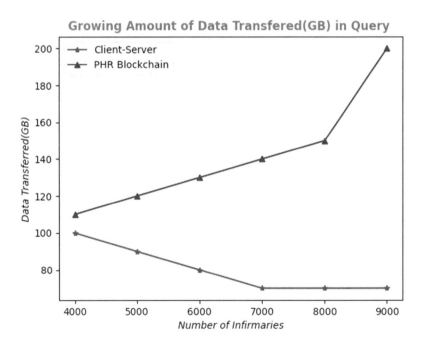

Fig. 13 Comparison of data transferred (GB) vs. growing number of infirmaries during record querying

Table 1 Execution time (hours) vs. growing PHR and infirmaries

Parameter	Factors	Average execution time (hours)			
		Updating the records		Querying the records	
		Client–Server	PHP blockchain	Client/server	PHP blockchain
Health records	+1500	2.05	14.56	2.05	0.5
Infirmary	+15	4.50	58.88	4.50	0.62

Table 2 Data Transferred (GB) vs. growing PHR and infirmaries

Parameter	Factors	Data transferred (GB)			
		Updating the records		Querying the records	
		Client–server	PHP blockchain	Client/server	PHP blockchain
Health records	+1500	26.72	260.62	26.72	29.01
Infirmary	+15	0	1045.88	0	10.35

for database update and demand. This indicates that with growing numbers of medical records and hospitals, PHR blockchain surpasses client/server for database questions.

Table 2 displays the quantity of information transmitted from client–server and PHR blockchain methods for database upgrade and question. In all application scenarios, the client/server method exceeds the PHR blockchain.

12   Conclusion

Although several researchers have researched the use of PHR blockchain for healthcare data processing, there is no assessment of this emerging approach with the conventional client–server model to our understanding. The customer–server model is suffering from data management, data duplication, accessibility, protection, and confidentiality concerns. Because of its integrity, stability, privacy, and data duplication characteristics, the PHR blockchain model has a strong capacity to improve record management. We provided a systematic study of the two frameworks for data processing in healthcare in this chapter. We have established a simple healthcare framework based on the PHR blockchain model to evaluate the efficiency of the models under review for upgrading and parsing medical records. The findings of the simulations show that the model of the PHR blockchain will result in important enhancements. In general, this is in a patient-centered method where patients and/or doctors regularly visit medical records to create a consistent outlook from various hospitals using deep learning algorithms for an improved evaluation or diagnosis. Our findings indicate that, in comparison to the client–server model with a growing number of medical records, the medical record query for the PHR blockchain framework is 12.0 times quicker. However, given the time complexity and the volume of data transmitted if the process requires a modification of the medical record, the PHR blockchain control system is more expensive than the client–server system model. This is attributable to the system of the agreement included in the modification to the database. Implementing the established platform to test its privacy and protection is one of the expected research areas.

References

1. Fusco A, Dicuonzo G, Dell'Atti V, Tatullo M (2020) Blockchain in healthcare: insights on COVID-19. Int J Environ Res Public Health 17(19):7167
2. Zhang P, Schmidt DC, White J, Lenz G (2018) Blockchain technology use cases in healthcare. In: Advances in computers, vol 111. Elsevier, pp 1–41
3. Agrawal R, Chatterjee JM, Kumar A, Rathore PS (2020) Blockchain technology and the internet of things: challenges and applications in bitcoin and security, Apple Academic Press. https://books.google.co.in/books?id=FCoMEAAAQBAJ
4. Ismail L, Materwala H (2020) Blockchain paradigm for healthcare: performance evaluation. Symmetry 12(8):1200
5. Roehrs A, da Costa CA, da Rosa Righi R, da Silva VF, Goldim JR, Schmidt DC (2019) Analyzing the performance of a blockchain-based personal health record implementation. J Biomed Inf 92:103140
6. Tandon A, Dhir A, Islam N, Mäntymäki M (2020) Blockchain in healthcare: A systematic literature review, synthesizing framework and future research agenda. Comput Ind 122:103290
7. Reda M, Kanga DB, Fatima T, Azouazi M (2020) Blockchain in health supply chain management: state of art challenges and opportunities. Procedia Comput Sci 175:706–709
8. Abu-Elezz I, Hassan A, Nazeemudeen A, Househ M, Abd-Alrazaq A (2020) The benefits and threats of blockchain technology in healthcare: A scoping review. Int J Med Inform 104246
9. Rajmohan R, Kumar TA, Pavithra M, & Sandhya SG (2020) Blockchain: Next-generation technology for industry 4.0. In Blockchain Technology (pp.177–198). CRC Press
10. Kalaipriya R, Devadharshini S, Rajmohan R, Pavithra M, Ananthkumar T (2020) Certain investigations on leveraging blockchain technology for developing electronic health records. In: 2020 International conference on system, computation, automation and networking (ICSCAN), pp. 1–5. IEEE
11. Kumar KS, Kumar TA, Radhamani, AS, Sundaresan S (2020) Blockchain Technology: An Insight into Architecture, Use Cases, and Its Applicationwith Industrial IoT and Big Data. In Blockchain Technology (pp. 23–42). CRC Press
12. Manju Bala P, Kayalvizhi J, Usharani S, Jayakumar D (2018) A decentralized file shareing & data transmission in peer to peer communication using edonkey protocol. Int J Pure Appl Math 119(14):1027–1032
13. Matilda S, Kumar, TA (2020) The Winning Combo: Cryptocurrency and Blockchain. In Blockchain Technology (pp. 199–217). CRC Press

Secure Vehicular Communication Using Blockchain Technology

N. Padmapriya, T. Ananth Kumar, R. Rajmohan, M. Pavithra, and P. Kanimozhi

1 Introduction

The number of autonomous and smart vehicles has been drastically increased in recent years. The characteristics of user groups in the VANET (vehicular ad hoc network) are focused on how successful the vehicle interaction is. The primary purpose of a vehicle network is to quickly disseminate details concerning life-threatening incidents, that is, to give timely traffic and accident reports. However, the transmission of crucial event data in a specific area in a complex VANET area and in the presence of malevolent automobiles is still a challenge. There are several security considerations regarding current VANETs [1]. Several critical messages cannot be disseminated correctly because there are inaccurate and unreliable data on malevolent vehicles. This also increases the injury risk to nearby drivers and pedestrians. As VANET technology encounters difficulties, one of its ongoing challenges is that in a VANET, driverless vehicles dynamically join or leave the network. The blockchain is increasingly gaining attention from researchers and has great potential in many areas [2]. We address the fundamental problem of information transmission in VANETs by using blockchain technology. In VANET, all vehicles can access the past event details of the vehicle, which can be used to control the ground reality of the vehicle information. For VANET applications, a conventional blockchain is not the ideal implementation. Hence, a new type of blockchain is designed, which is feasible for the use of VANET event messages and transactions. The blockchain technology solves the foremost problems that current VANETs face and provide protection for the diffusion of sensitive information [3]. New blocks are created in our scheme focused on event messages close to Bitcoin transactions, and hashes of successive blocks are sequentially connected together to create a blockchain. This chapter aims to research how to use blockchain technology in VANETs

N. Padmapriya (✉) · T. Ananth Kumar · R. Rajmohan · M. Pavithra · P. Kanimozhi
IFET College of Engineering, Villupuram, Tamilnadu, India

© Springer Nature Switzerland AG 2022
P. Raj et al. (eds.), *Blockchain, Artificial Intelligence, and the Internet of Things,*
EAI/Springer Innovations in Communication and Computing,
https://doi.org/10.1007/978-3-030-77637-4_8

to distribute trustworthy event messages securely. In order to increase the interoperability and timeliness of the VANET message distribution, blockchain is based on the local area, which is not dependent on other nations. We assume a public blockchain that manages all node resilience and message reliability in a given nation independently and stores them [4]. Various methods of blockchain are presented based on the public and private blockchain. The mechanism of consensus plays an important role in evaluating a blockchain's security and scalability. We will concentrate on the consensus process for proof of work (PoW), which is a good and proven protection that is suitable for a public blockchain [5]. In addition, the propagation delay of the block is decreased by the usage. Our chapter's main contributions can be summarized below:

(a) For storing the message and integrity in the VANET, we suggest a blockchain scheme. In this method, the integrity of the node and message functions like Bitcoin blockchain transactions that provide other vehicles with situation on the ground.
(b) A local blockchain is used; focused on environmental areas and distinct of chains from various nations, we are seeking to boost the interoperability of the blockchain.
(c) We intend to eliminate block generation latency by incorporating edge computing as a technology standpoint into the VANET blockchain. By unloading the complicated computation to the edge computers, edge computing would decrease latency, thereby providing real-time VANET applications.

The remaining part of the chapter is arranged as follows: VANET fundamentals are defined in Sect. 2. A brief overview of the trust models that are used in VANET is given in Sect. 3. It also provides the history of blockchain and briefs various forms of consensus mechanisms that are used in the present blockchain. The theoretical description of blockchain and VANET is explained in Sect. 4. Section 5 introduces a new form of blockchain, and the performances of the proposed blockchain-based VANET for the dissemination of protected messages are discussed in Sect. 6. In Sect. 7, the potential outlook of the blockchain-based VANET is provided. Finally, the chapter is concluded in Sect. 8.

2 Fundamentals of VANET

This section offers a theoretical description of the underlying principles and context details relating to the VANET. VANET has recently played a vital part in saving the lives and characteristics of users through the dissemination of vital incident knowledge through the advancement of vehicle technology. Two forms of communication exist within VANET: vehicle-to-infrastructure (V2I) and vehicle-to-vehicle (V2V) communication. Vehicle-to-everything (V2X) is further common these days, where all reflects pedestrians, cyclists, and everything that interact with the vehicle. The automobiles connect with road side units (RSUs) that are mounted beside the two sides of a road in case of Vehicle-to-Vehicle communication. A system for the security analysis of alert messages disseminated randomly between vehicles available in a VANET has been developed by Obaidat et al. [5]. Malicious vehicles have commonly been spread within this network. This work measured the likelihood that a vehicle will correctly receive a message at a given distance. In addition, the correctness of the transmitted communications was enhanced to some degree through the fusion of all the messages received. Raw et al. [6] suggested a "receiver consensus" broadcasting method for the delivery of caution messages in a vehicle. This algorithm used environmental data to support the nodes in reaching a consensus on a forwarding node option. The neighboring vehicles were autonomously rated and given priority to broadcast the message, based on their distance from the ideal spot. Hamdi et al. [7] suggested a simulation model based on the agent that applies the self-pruning transmission procedure to the vehicles' disseminated threatening messages. A roadside tracker for transmitting warning messages was used in this algorithm. The messages were sent with up to a certain number of hops to all surrounding vehicles. Kamel et al. [8] suggested an obliging emergency braking alert scheme using a dedicated short-range communication protocol (DSRC) along with a camera sensor to resolve the location accuracy and time-critical issues in the delivery of caution messages between vehicles. In addition to the method of reducing message transmission, the condition of neighboring cars was created. The protection and authentication of the distributed messages was, however, overlooked by this work. There was a very strong probability of a message coming from a malicious vehicle.

Fang et al. [9] reported that their system can be authenticated through a two-stage approach. A Bayesian network was used, which analyzed several environmental variables alongside their correlations. A density-aware dissemination protocol was proposed by Selvi et al. [10] to ensure, in particular areas, timely dissemination of disaster alert information. A protocol for VANETs (DRIVE) was proposed by Villas et al. [11] to solve the broadcast storm issue. Pattanayak et al. [12] have developed a reputation scheme for VANETs to mitigate Sybil attacks. This logo was used by RSU in its certificate-granting range. A means of mitigation against black hole attacks was proposed by Tobin and Thorpe [13]. The extent to the multistep security approach included detecting attacks, accusing nodes, and blacklisting malicious ones. Alsulaim et al. [14] proposed an algorithm to detect denial-of-service (DoS) attacks in a vehicular ad hoc network. An unknown sender was detected by the device reading high-velocity packets at a high rate [15]. However, this algorithm consistently failed to distinguish numerous malicious bits of data that were concurrently being transmitted. The vulnerability characteristics of the VANET were briefly described by Malhi et al. [16]. Network size, heterogeneity, uncertainty, and the use of wireless connections can be examined. This work identifies the various security attacks on VANET relative to each layer of authentication. Countermeasures were also thoroughly investigated for each assault. The intrusion detection is identified by the clustering algorithm in VANET; AECFV was proposed by Zhou et al. [17]. For perceiving attacks such as a wormhole, black hole, Sybil attacks, and packet replication, the writers used detection and categorization techniques. In order to prevent communication and computing blockages in the reversal of new automobiles, a stable identity-based system with aggregate signatures was proposed [19]. A certified reputation framework was proposed by Abassi et al. [20], in which vehicle reputations were revised regularly by centralized authorities. The vehicle's credibility level was calculated by the conduct of a vehicle over the network of Manivannan et al. [18]. Typically, contact between vehicles is wireless, which is susceptible to interception and hacking. In addition, this coordination in the event of natural disasters resulting in severe traffic congestion is often inaccessible or failed. When a large number of vehicles interact with one another, the RSUs have their limits on storing and processing it [21, 22]. The centralized structure is also very vulnerable to attack, too. In addition, the transmission rate of these messages is restricted to a particular geographical location [27].

3   Trust Model for VANET

The present trust models can be divided into three key categories: through a trust circle, based on a social circle, and trust one particular individual shown in Fig. 1. It includes the entities and how their trust is modeled, as well as being data-based, and also includes a hybrid model. The entity-based trust models evaluate the reliability of all parties involved in the transaction to determine if the transaction is secure. In the articles [24–26], the authors presented more structured approaches for verifying the reliability of nodes in networks. Network nodes are not trustworthy enemies all of the time, and they may say things that don't always agree with you. Because of their high mobility, most vehicle nodes do not have a constant position relative to the deployment location, making it difficult to determine their trustworthiness. In a similar trust model, called "data-centric trust," the trustworthiness of the vehicle is taken into account. Also, "data-centric trust" puts trust in one and another instead of in the node itself. The authors [4, 25] used a Bayesian inference decision module to calculate how "real" one's reported event was. The inference module is based on previous probabilities. It also needs instantaneous communication between the nodes, and the topology of each node is dynamic. The vehicle communicating with the vehicle that will be discussing the value of the van may be manipulated by the malicious vehicles. Additionally, since the trustworthiness of the vehicle that is

Fig. 1 VANET-based trust models

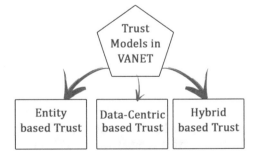

sending the message is not ensured, the trustworthiness of the message itself may be compromised. Therefore, the hybrid trust framework was designed, and it was used to assess reliability by combining entity-based trust and data-based trust. In the work [26], the authors developed a trust management mechanism that evaluates data reliability through the number of messages they receive from different Internet nodes. The trustworthiness of the network is judged by the number of recommendations and functional connections it has. Under normal or near-normal situations, the systems under consideration do not consider the amount of data available. Thus, we are working to overcome the inherent disadvantages of our present hybrid trust system by improving the way we are verifying the trustworthiness of messages and nodes.

We use a blockchain platform to create a trust level for the VANET and then calibrate the trust level with subsequent messaging interactions in order to achieve a level of trust within the VANET system. By allowing nodes to store trust levels on the blockchain, we can quickly store the trust levels for each node and node responses in a distributed database. We offer enhanced security and privacy by using the blockchain.

4 Blockchain and VANET

This section delves into the underlying concepts involved with the VANET, the blockchain, and all of the different components that go into a VANET. The basic components and trust models of the VANET are listed out. We compare the characteristics of the various cryptocurrency networks using a consensus mechanism. Vehicles ad hoc (VANETs) arose as a subset of the mobile ad hoc network (MANET) application [28]. The vehicles are actually a network of small and mobile robots with an onboard transceiver, which are mobile enough to dynamically join and leave a wide variety of ad hoc networks with little or no assistance. As a proven solution to intelligent transportation systems (ITS), VANET has applied techniques to intelligent transportation systems (ITS). Recently, social networks have become the subjects of research by scholars in the wireless mobile communication field and the VANET. The purpose of VANET is to inform drivers of traffic delays via roadside units, road signals, and traffic cameras and to promote road safety by providing accurate and timely road information in a well-defined area, increasing road safety and road traffic performance, and improving road users' information capabilities. When VANET is used, vehicles are prioritized in a network and are used as network nodes.

VANETs will provide vehicles with inter-vehicle communication enabling vehicle-to-vehicle and vehicle-to-infrastructure communication. In the vehicle-to-vehicle and vehicle-to-infrastructure paradigm, it is essential for the vehicles to have a dedicated short-range communication unit and a transceiver unit. VANET will provide convenience to the drivers and the passengers and will also provide navigation. Vehicle traffic engineering is composed of both safe signals and other nonhazardous signals such as speed of vehicles and camber and direction of vehicles (value-added comfort application). VANET also comes with its challenges, which we need to overcome, especially the security issues that we need to deal with. As the road starts to get busy, with cars and trucks entering and exiting the highway in close succession, the driver has to gather information on which route to take, what roads and traffic conditions to expect. In order to get this information to the proper individuals at the proper time, it is necessary that it be delivered in a timely fashion. They will reach the destination safely. Despite the current amendments, some malicious nodes obstruct the required safety messages before transmission to the user, resulting in extended delays and serious injuries. The research [28] shows that a VANET (virtual anonymous network) is a network where two people can interact by sending small messages to one another, and it may be useful for individuals to communicate with one another online. This network must have characteristics that are different from other wireless communication networks, such as its high mobility and its volatility, and those are the characteristics that it has been susceptible to attacks. VANET can be characterized by a relatively decentralized structure and dynamic topology. Therefore, it is important that the security of vehicles, users, and data is understood because the identification of malicious parties, faulty actions, and incorrect information has become difficult. VANETs where you can interact with your surroundings are adopting new applications with the new tech, like the autonomous VANET where you can intersect with your surroundings. Although it is easy to use social media to communicate important information that benefits the users, we run the risk of fictitious attacks that can hurt people due to the random nature of social media. In order to create a feasible network to have messages to be validated, the vehicles that pass by the network need to be tracked, so a picture of where they are and what they are doing at any time is made. Because there are so many laws, regulations, and rules that essentially protect normal users, there needs to be a balance between the beneficial needs of users and the need for the governing powers to be able to access to their data that they hold. Several researchers have developed different methods for preserving privacy from violations of HIPAA, such as pseudonyms [29] and anonymous authentication [30], which could achieve the objective of preserving the privacy of their users, as long as they are unable to link pseudonyms to the user. Nonetheless, we would recommend that the pseudonym scheme be used only when high security is required for privacy because the scheme could not be used to link pseudonyms to users, which could be damaged by traffic information. Not only might the group of cars lose their proper assignments when the fake vehicle association is incorrectly verified, but also the fake vehicles might be recognized as valid ones by such a group as long as there are a number of invalid vehicles. There is a variety of trust management and privacy protection options available for VANETs, but issues remain regarding security and trust for vehicles [23]. Therefore, to ensure security, privacy, and trust, it is important to make certain that design networks have both strong security and privacy, as well as mutual trust among network participants.

Vehicle ad hoc networks are systems based on mobile ad hoc network (MANET) techniques. Many experts accept VANET as an intelligent transport system. VANET has captured the attention of researchers in networking and signal processing. The goal of VANET is to improve traffic flow by providing more accurate traffic information and improving traffic safety through faster communication. In VANET, vehicles move within a network and are nodes on the network. Vehicular communication is either vehicle-to-vehicle (V2V) or vehicle-to-infrastructure (V2I). Onboard units (OBUs) and roadside units (RSUs) in VANET establish one-hop communication between each other through dedicated short-range communication (DSRC). VANET provides various services to drivers and travelers. There is information on speed/speed-related accidents and curves and bankings. Safety information is given higher priority than alerting relevant data because it notifies drivers of coming hazards that enable them to adapt immediately. VANET faces certain challenges, including health and data privacy. Vehicle operators on roads are required to provide meaningful information to drivers on road conditions such as traffic congestion.

VANET focuses on transportation and uses smart sensors and wireless interconnection. As one of intelligent transportation systems' main components, VANET enables vehicle communication, allows vehicles to communicate with one another, and offers a mechanism to help manage traffic. The VANET provides the driver with the ability to make the right decision by reviewing the traffic conditions beforehand. Figure 2 depicts the basic architecture of VANETs. The backbone network handles

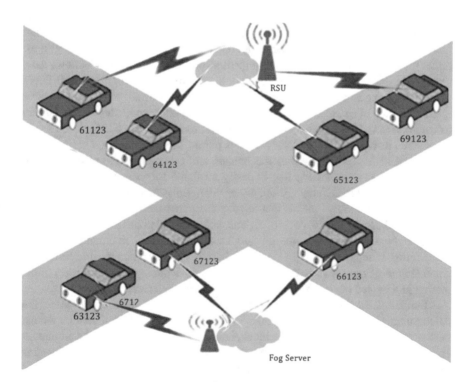

Fig. 2 Basic vehicle ad hoc network (VANET) communication

the core networking. The wireless ad hoc networking is used for private wireless networks. The onboard units are integrated into the vehicles, assigned the responsibility of information storage. OBUs also send commands in vehicle-to-vehicle and vehicle-to-roadway networks. Vehicles also have a multiapplication unit (MAU) or single application unit (SAU), allowing the vehicle to use the provider's applications. There are three types of communication between VANETs: roadside, ad hoc, and vehicles. Recursive sheepdog includes the Internet, gateways, and RSUs. RSD is susceptible to certain types of attacks. Hackers will send several emails to VANETs trying to overload the RSU. Smart vehicle devices are AHD related. AHD must defend against attacks, including routers and authentication. To forge ID cards, attackers generate fake original cards. Interaction device allows interaction between SAU or MAU and OBUs. Insider threats consist of individuals or groups attempting to gain unauthorized access to data in the VANET. Sharing and communicating with peers and colleagues is an essential part of the VANET, so efficient and trustworthy data sharing is very important. Since these VANETs are exposed to serious threats, security and privacy issues must be addressed. VANET, in all its various domains, is equally important and needs to be monitored for malicious attacks. VANET security issues are an issue due to all of the following reasons. Attackers gravitate toward VANETs because the data they store, transmit, and collect are sensitive.

"Decentralization" means that blockchain is decentralized since it's composed of numerous nodes. There is no equipment and/or management organization for the equipment [35]. The mathematical algorithms for verifying, accounting, storing, maintaining, and transmitting blockchain data are set up by decentralized systems, rather than being controlled by central institutions. "Decentralization" allows data, assets, and information to be exchanged or shared freely. Distributed is a different concept than decentralized. Some uses of ICTs may be centralized. A central server distributes data in a distributed network enabling parallel processing of the data. In a decentralized system, there is no central server to store the data [36].

All data information of the blockchain is public, thereby making it easily accessible to all nodes. The consensus mechanism and rules set by the blockchain are verified by open-source code that is easily available for inspection. Anyone can participate in the blockchain system, either via permission or without the need for permission. "Autonomy" means that anyone can participate in the blockchain network, providing a complete copy of all transactions to all nodes participating in the network. Nodes store a common blockchain, which is maintained through competition-based computing. Consensus-based specifications are used by blockchain technology to securely enable all nodes in the entire system to transfer information in a trusted environment, regardless of human intervention.

"Tamper resistance"means that a single or even multiple nodes cannot affect the database unless they control more than 51% of the nodes, and even then they must be controlled at the same time. The blockchain uses a hash function and asymmetric encryption to ensure that information will not be tampered with. Since each block is linked with the previous block by a cryptographic proof, any block can be modified to modify a transaction in a previous block. The cryptographic proofs of all previous blocks are reconstructed, making it essentially impossible to tamper with the blockchain. Because nodes have a fixed set of information, the real identity is protected and one can be "anonymized." The blockchain breaks the chain by determining whether or not a transaction is valid. Therefore, neither party needs to reveal the identity of the other.

The most important aspect of blockchain technology is its decentralized nature. Blockchains are a collection of cryptographic techniques that allow peer-to-peer transactions to occur without third parties. As a result, blockchain technology has become a vital part of digital currency systems. We have developed a secure distributed network that combines a variety of different frameworks. The blockchain is best equipped for this research exercise because it allows the analysis of the collected data to be done quickly, efficiently, and accurately. In the further reading of the chapter, we shall discover more of the details of the blockchain and how it works. In order to overcome the centralized network structure of Bitcoin, Satoshi Nakamoto proposed a decentralization and consistency of a database system called a blockchain. A "distributed ledger"is a "chained ledger," where the output of one transaction contains the inputs for the next one.

Each ledger is a peer-to-peer distributed ledger consists of multiple books that store transactions. Each block contains a record of the previous block's hash prior to that block's validation. Each money payment is digitally signed by the sender and the recipient. Other blockchain services include a trustless distributed network that verifies/validates transactions. There are some differences between how blockchains store information compared with how data are stored in MDA. However, blockchain technology is enabling many of the features of MDA and might be used for MDA. The fundamental ideas behind blockchain are that of a genesis block, which is the very first block in the blockchain, as shown in Fig. 3. The block (or, chain) of information contains information that is passed on to all other blocks by all other nodes in the networking system. As the first step in this process, the block is composed of saved information encoded as a hash. The block also contains a list of data about the previous block.

As Fig. 4 illustrates, block numbering starts at zero and the block header is the hash of a previous block. A block holder keeps a portion of blocks, which include lists of transactions, as well as additional data, depending on the need of the respective blockchain.

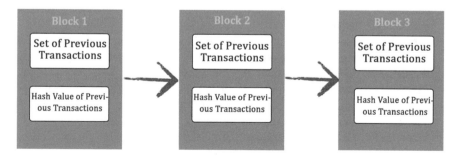

Fig. 3 Architecture of blockchain

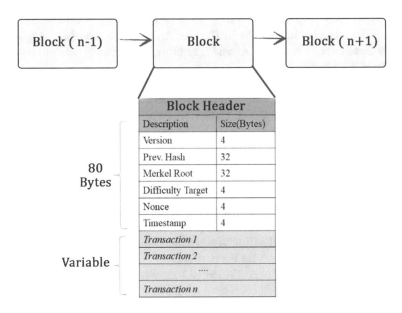

Fig. 4 Block header format

4.1 Major Features of Blockchain

Some of the salient characteristics of the blockchain include the following:

- Immutability: Blockchain characteristics include the inability to change information. Once a piece of information is written in the blockchain, it cannot be changed or deleted. Furthermore, information cannot be added at will.
- Distributed and trustless environment: In the blockchain, any node can synchronize the entire blockchain and validate its contents without central control. This approach's advantage is that it allows the data to be on a single server, offline. It helps build trust in a trustless environment.
- Privacy and anonymity: It can be maintained by using the blockchain capability and cryptography topics. People join anonymously. For example, other people can't know user information itself. It ensures that personal information is private, secure, and anonymous.
- Faster transactions: It is easy to set up the blockchain; one of the features of which is that transactions are confirmed quickly. You are inputting or processing transactions within a few seconds or a few minutes.
- Reliable/accurate data: The blockchain has distributed data, so it is reliable, accurate, consistent, timely, easily shared, and open in nature. It is very robust. It can survive malicious attacks as well as a single point of failure.
- Transparency: It is full transparency of all the transactions that happen in the network. Anyone can view the transactions as they happen.

5 Proposed Blockchain-Based VANET Communication

Many academic researchers have been brought in by the promise of distributed ledger decentralization's benefits that can be derived in the scientific field as a whole. While blockchain is consisting of an infinite number of nodes, it is important to note that the blocks don't necessarily have to contain the same data that are connected in a nonreversible sequence to form the blockchain. Because this information technology is growing increasingly successful in different disciplines, it has gained the attention of a range of key players in releasing crucial and reliable information to benefit us all. The fundamental assistance of distributed ledger is called Bitcoin cryptocurrency, a decentralized, distributed, resilient, secure computing paradigm that will allow applications to be transparent over a P2P network and allow people to manage information without the need for a central authority. In the VANET environment, this technology is a core that helps to support the management of the underlying data and reality. As just a huge portion still, developers agree that automobiles can share the preceding event list and its data if it is positioned in the blockchain network.

The framework is a method of making the truthfulness of access point and message passing in VANETs (vehicular ad hoc networks) to be assured by the fact that they are placed on a public blockchain (distributed shared ledger) as underlying data for other VANETs. Sending bulk transactions of donation money from donors to the retailer on a blockchain will not be enough of a technology for the VANET. In order to fit the VANET user, the digital currency used to pay for goods has to be implemented and supported by all participants in the network. The possible explanation for the variation made to the original transaction is to guarantee the adequacy of features for the VANET system, as well as the solution of ensuring protection for valuable data transmission and addressing the VANET issues. The difference method adds new blocks to the blockchain by adding new data as events, comparable to transactions in Bitcoins, besides the hash function sequences of the blocks that are to be connected to the blocks in the blockchain in sequential order. This system is maintained by an intricate distributed algorithm. Effects on reliability and timeliness are ensured, while scalability is ensured by using self-contained independent geographic-separated blockchains. A public blockchain is a network technology that is used to access and handle the node and data integrity that are given in a geographic location. Depending on the type of public ledger, that is, either public or private, a wide variety of smart contracts are to choose from at the onset of a blockchain. Because of the feedback loop between consensus and security and scalability, the safety and interoperability of the blockchain are strongly dependent on the consensus protocol used. One particular blockchain system isn't appropriate for the VANET problems. But the solution they came up with didn't work; it became necessary to develop a new, improvised blockchain. The information contained in each block is comprised of vehicle Vi value, the identity of the vehicle who transmitted the message, the Mi value of the message, the RSS_i value of the vehicle, the value of the timestamp at the moment the message was transmitted, and the hash value of the transaction. Event messages are both sent to and signed by the individual sending the message. The event messages will then be posted on a public chain, and this will give rise to the reliability of the event message's ability.

The problem of confidentiality in the VANET can only be solved with a simple, public, blockchain-based distributed ledger. Because it is informally, a sort of mechanism is introduced that facilitates feature adaptation to counter the errors of each stage of the blockchain. Each block is made up of a vehicle's identity, the message from that vehicle, the comparative signal quality and indicator quality of that vehicle, the date stamp of that vehicle, the hash value of that vehicle (hash), and the other transaction entity's root value (root). In this VANET, when an event message is issued from a node, it is relayed only through nodes that have a copy of a valid, signed, and publicly accessible event message.

5.1   Blockchain-Based VANET Architecture

Distributed ledger technology is built at the MAC (media access control) layer in the suggested method to achieve secure and reliable information exchange throughout the network. In addition, the preferred method also incorporates a signature-less public blockchain so that the information that is shared cannot be had by an unauthorized party.

To support the big data size, 5G technologies are able to maintain the VANET provider with both the execution of the change architecture. The performance measures state that, with the help of 5G technology, they can attain an information rate of 10 Gb/s and less than 1 ms of latency. Besides that, the four kinds of use cases defined in VANET/smart VANET, respectively, are well-characterized key innovations with greater promise to provide guaranteed execution of VANET's tasks, like roaming, sensor collection and transmission, data dredging, and list-sharing. And they are probably the only areas of L2 IoT that the IoT industry had been treating seriously—especially the category of smart VANET, which is expected to see rapid growth over the next decade. And in a possible framework, communication between two vehicles results from a system design that positions the home agent as a driverside vehicle on a side road. The device architecture of the proposed system is shown in Fig. 5, in which an interaction between two vehicles occurs. In this study, 25 distinct blockchains have been viewed for the implementation of the product certificate chain (Pr_Cer_Ch), revocation chain (Re_Ch), message blockchain (Mess_Ch), and trust chain bridge (Tr_Ch). Since this work uses a cryptocurrency, it could be prescribed by government entities, including the law enforcement agency (LEA) and security certification authority (CA). A standardized protocol, where all participating vehicles adhere to certain regulations, leads to a need to have the system create trust values, which will be used in the system to prevent double spending while also giving the government the ability to collect tax.

In the proposed solution, there are six phases: system initialization, system authorization, generation of message ratings, estimation of trust value offset, miner

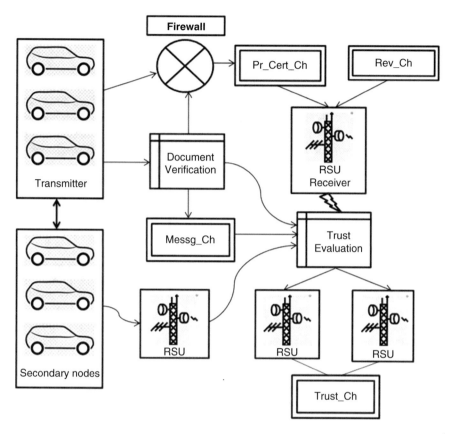

Fig. 5 Blockchain-based framework in VANET

Fig. 6 Data flow of the
proposed system

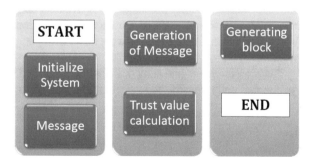

election and block generation, and distributed consensus, as shown in Fig. 6. During
the first phase, a system is initialized to start the process of being fixed. Thus, the
whole process is responsible for verifying the identity of the nodes and issuing a
certificate to them, when a switch moves in and out of a network. The actual authentication process is a follow-up process to the first phase, where the nodes assume a

root position or authority over the profile, as well as being the initial root. The next phase works with every other node to verify the security level against their peer and verify that they are allowed to communicate with the specified peer. The expanding email problem provides a rating on the messages sent by the communicating nodes. Following this, the trust level of each node is assessed, based on information that is already being collected about that node and on the programming of the node itself. Next, the node then elects a representative to act as a miner. After that, it performs a system of blocks, which implements the blockchain technology for tracking of the nodes in the system. The last phase will be the distribution of consensus, so everyone can be informed and the network can function properly.

6   Experimental Simulation and Analysis

In this study, the research is investigated by analyzing the effect of denial-of-service on the blockchain-based vehicular network. These modern techniques have been contrasted with other cryptographic algorithms based on smart cards (SA), contract procedure (CP), and privacy-based authorization (PBA) [31]. Veins tool [32] is selected as the visualization tool for this analysis because it had all the functionality we needed: interactive parameter settings and tracking of automobiles in response to a controller, a practical chart, and transportation paradigm. Veins could even interlink between automobiles on the fly. A composite model integrating the capabilities of the channel modeling methodology with those of the road-based simulations was selected, in particular, to provide the multimodal synchronized computation. Veins functions as an object-oriented network emulator and SUMO [32] for cities model. Traci combines both supervisory control and data collection (SCADA) and object management community with the transportable instrumentation and calculation interface standard. A live data contact between the motor vehicles emulation framework and the network emulation system can be achieved in Veins. The channel modeling software has a major impact on the traffic prediction program.

The measurements and statistics used by the SUMO tool are as seen in Table 1. As for the node density, top acceleration, angular velocity, peak displacement,

Table 1 SUMO tool attributes

Attributes	Measurements
Node density	200
Top acceleration	54 km/h
Angular velocity	35.892 r/m
Peak displacement	71.469 r/m
Motor size	43 in.
Motor distance	3.892 in.
Rider abnormality	0.3

Table 2 OpNeT tool
specifications

Attributes	Measurements
Duration	100 min
Track length	10 km
Max_speed	120 km/h
Data_rate	10 mbps
Congestion_window	35 m
Motor_sense	−90 db
Reconfigure_limit	$0.01 * 10^1$ m

motor size, motor distance, and rider abnormality are configured as the factor of the control system.

Table 2 lists the specifications for the OpNeT++ emulator [32] such as duration, track length, max_speed, data_rate, congestion_window, motor_sense, and reconfigure_limit.

The efficiency of the implementation in VANET was analyzed employing the node distribution scale, end-to-end latency, packet drop, and packet overload. Any of the computing metrics were tested with DoS attack. Therefore, the average and standard deviation (SD) parameters were estimated using the node data rate (NDR) [33]. It corresponds to the percentage of packets that effectively arrived at their destination. Interconnection formula is derived by calculating the proportion of the amount of data chunks transmitted, c^t, to the amount of data chunks received, c^r, in the system, as demonstrated in (1):

$$NDR = \frac{c^r}{c^t} * 100\%$$

(1)

In Fig. 7, the suggested strategy preserves NDR at an acceptable rate by rejecting the Ip spoofing attacks, and the system continued to operate without interruption. The suggested approach culminated in an estimated NDR of 0.82, a SD of 0.034267, and a CI of 0.018604 with 200 linked routers in the system. There was a significant issue with the suggested and modified specifications as it wasn't prepared to defend the device from the Ip spoofing attack.

Edge latency [32] is a period required for a message to appear at the target from the transmitter. NDS suffered substantial influence from the edge latency encountered by the system. The point latency is calculated by measuring the difference in period between the delivery of a payload at its target automobile and the sending of a payload by the same automobile. Equation (2) is used to measure point latency, where ED is edge delay, A_{ti} is arrival time, and S_{ti} is sent time.

$$ED = \sum A_{ti} - S_{ti}$$

(2)

From Fig. 8, it is clear that the suggested approach established a stable latency, with an interruption of 0.31 s, 0.45 s, and 0.52 s, although there was an Ip spoofing attack in the system as compared to SA, CP, and PBA, collectively. Whenever a

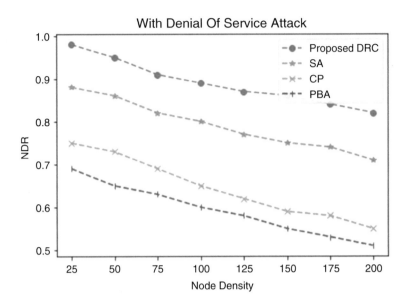

Fig. 7 NDR performance analysis

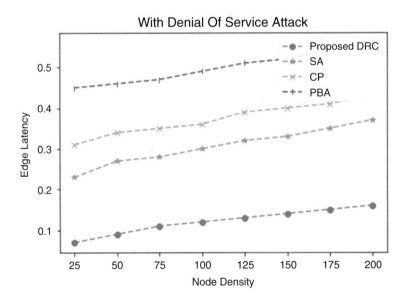

Fig. 8 ED performance analysis

quantitative study was done, the preferred method delivered a median edge delay of 09 ns with a normal distribution of 1.32 ns and a correlation coefficient of 1.57 ns. The measurements were appropriate in the suggested system by using the flexible ledger genre. They are connected using previously hashed message blocks, which seems to be an important feature of the suggested technique.

Packet loss percentage [33] is the percentage of messages that are not delivered to the recipient in the system. It is determined by the proportion of bytes dropped Bd, to bytes sent, Bs, in the system.

$$PLP = \frac{B^d}{B^s} * 100\%$$

(3)

As can be seen from Fig. 9, the preferred method preserved information depletion at a manageable amount with a gap of 19.8%, 15.6%, and 16.8% when we targeted the channel with an Ip spoofing attack. Regarding quantitative research, the cumulative message drop was 23.8%, the normal distribution was 1.97, and the correlation coefficient was 1.47. For this reason, the data loss of the standard implementations was far higher than that of the suggested system from the start, causing a great amount of message deletion.

Communication impedance (CI) [33] applies to the amount of packets sent across the wireless medium by the overall size of the message. Larger information packages hold multiple pieces of information, which ensures that they would be larger and take further processing time (Fig. 10).

$$CI = P_{overhead} / P_{total}$$

(4)

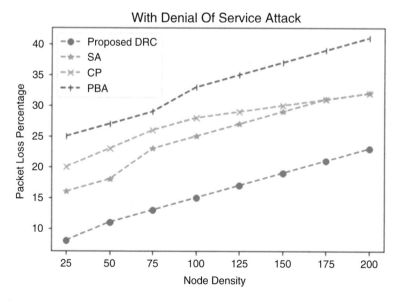

Fig. 9 PLP performance analysis

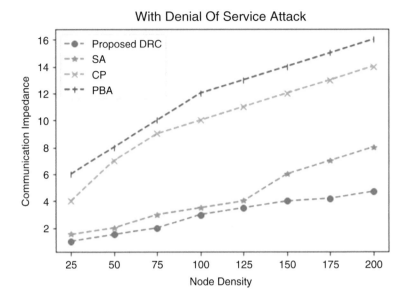

Fig. 10 CI performance analysis

The suggested approach experienced a significant rise in communication impedance as opposed to CP and PBA, with a disparity of 13.8 kilobytes and 6.9 kilobytes, collectively. The total communication impedance of all channels is 4.5
kilobytes with a normal distribution of 0.18 kbps and a statistical significance of
0.12 kilobytes. The preferred approach experienced a substantially lower latency for
handling payments than the standard procedures owing to the usage of a cryptosystem. According to the implementation analysis, the proposed model outperforms
the baseline techniques in guidelines of protection as a priority. The beauty of ledger
techniques is that if a piece of evidence is applied to the database, it cannot be
deleted once it is registered. Furthermore, no new details can be applied to a database, which is a special and essential aspect of smart contracts.

Second, anybody can make changes or adjustments to the details stored on the
public ledger. A specific source of vulnerability wasn't feasible for this framework
because it was a highly sensible platform. Even the individual will be a member of
the system despite exposing their identities. We would claim that the suggested
framework has the strength of secrecy, honesty, reliability, and nonrepudiation. Rsa
algorithm authentication can be used in vehicle and infrastructure communication
forms of messages to ensure the privacy of messages and deter parody and message
manipulation attacks. From a public service honesty viewpoint, each message
incorporates a ciphertext and a timeframe. Each chain is bound to the prior frame,
thereby rendering spoofing unlikely. The trust evaluation protocol will enable only
valid peers to execute their tasks and disallow unauthorized users from disturbing
the system. In comparison, the distributed system of VANET will circumvent the
domination of central networks and can be avoided.

7 Research and Challenges in Ledger-Based VANET Communication

In an attempt to appease the increased reliability and limited bandwidth criteria for VANET-based IoT, it is vital to incorporate a decentralized system. This has accelerated the rise of distributed ledger technology [34], generating a vast range of potential exploratory areas.

- Introduce blockchain platform to address the diverse evolving vehicular IoT services. Autonomous driving mechanisms typically need hyper bandwidth, although automobile big data analysis has tighter specifications for processing platforms and storage space. Ledger architecture should understand the various implementations of blockchain. A further initiative is needed to come up with IoT use cases for blockchain.
- Distributed ledger for shared intellect: There is a vast range of computers, cars, detectors, networking systems, and compute clusters in the field of industry and shipping. There will be an increase in bandwidth and technical restrictions on potential flow. Edge analytics has gained a great deal of focus in modern decades due to its ability to boost mutual understanding. The traditional deep learning system focuses on a single platform to receive input from multiple customers. Conversely, in certain cases, there is no unified monitoring panel for cars. Blockchain can reduce processing costs by eliminating central servers. Research that included how to enhance intellectual ability in a distributed network by using augmented reality is a fascinating subject.
- Performance analysis of network using distributed ledger: A blockchain technology requires a vast amount of information to document a full history. A restriction in throughput is a serious problem in the ledger, particularly where autonomous vehicles are in service. Although the ledger has the benefit of decentralization, it is more costly to secure than the traditional method. The fundamental issue among federalization and network latency is to be addressed. The analysis of this vendetta is difficult for lack of research. Communication and computational technology combined with a public ledger would be necessary to build efficient and enhanced vehicular traffic networks.
- The efficiency of a ledger process is controlled by the distribution of interaction, processing, and disk usage. It is the dynamic essence of resource distribution in a decentralized distributed system. The motor vehicle ecosystems pose new problems for the production scheduling policies due to their diverse topologies. The connectivity weakness in a distributed network will make it difficult to communicate. The analytical efforts might not be concentrated sufficiently in one car. It would be helpful if we ensure effective scaling on a large scale.

8 Conclusion

To provide secure communication, VANETs have received tremendous interest from both researchers and the automotive industry. Fortunately, VANETs are still open to trust management for automobiles. Therefore, we proposed secure blockchain architecture in this chapter. This chapter proposes a safe blockchain architecture based on trust to effectively mitigate many network attacks while protecting VANET users' privacy and protection. This framework delivers warnings to avoid collision through the decentralized database (blockchain). We have used blockchain to validate the recipient of an alert message via the integrity of each vehicle. The proposed model has a high degree of precision in terms of V2V communication.

References

1. Martuscelli G, Boukerche A, Bellavista P (2013) Discovering traffic congestion along routes of interest using VANETs. In: 2013 IEEE global communications conference (GLOBECOM) (pp. 528–533). IEEE, Dec 2013
2. Nakamoto S, Bitcoin A (2008) A peer-to-peer electronic cash system. Bitcoin. URL: https://bitcoin.org/bitcoin.pdf, 4
3. Morgan YL (2010) Notes on DSRC & WAVE standards suite: its architecture, design, and characteristics. IEEE Commun Surv Tutorials 12(4):504–518
4. Raya M, Papadimitratos P, Gligor VD, Hubaux JP (2008) On data-centric trust establishment in ephemeral ad hoc networks. In: IEEE INFOCOM 2008-the 27th conference on computer communications (pp. 1238–1246). IEEE, Apr 2008
5. Obaidat M, Khodjaeva M, Holst J, Zid MB (2020) Security and privacy challenges in vehicular ad hoc networks. In: Connected vehicles in the internet of things. Springer, Cham, pp 223–251
6. Raw RS, Kumar M, Singh N (2021) Software-defined vehicular adhoc network: a theoretical approach. In: Cloud-based big data analytics in vehicular ad-hoc networks. IGI Global, pp 141–164
7. Hamdi MM, Audah L, Rashid SA, Mohammed AH, Alani S, Mustafa AS (2020) A review of applications, characteristics and challenges in vehicular ad hoc networks (VANETs). In 2020 international congress on human-computer interaction, optimization and robotic applications (HORA) (pp. 1–7). IEEE, June 2020
8. Kamel J, Ansari MR, Petit J, Kaiser A, Jemaa IB, Urien P (2020) Simulation framework for misbehavior detection in vehicular networks. IEEE Trans Veh Technol 69(6):6631–6643
9. Fang W, Zhang W, Liu Y, Yang W, Gao Z (2020) BTDS: Bayesian-based trust decision scheme for intelligent connected vehicles in VANETs. Trans Emerg Telecommun Technol 31(12):e3879
10. Selvi M, Ramakrishnan B (2020) Lion optimization algorithm (LOA)-based reliable emergency message broadcasting system in VANET. Soft Comput 24(14):10415–10432
11. Villas LA, Boukerche A, Maia G, Pazzi RW, Loureiro AA (2014) Drive: an efficient and robust data dissemination protocol for highway and urban vehicular ad hoc networks. Comput Netw 75:381–394
12. Pattanayak BK, Pattnaik O, Pani S (2021) Dealing with Sybil attack in VANET. In: Intelligent and cloud computing. Springer, Singapore, pp 471–480
13. Tobin J, Thorpe C, Murphy L (2017) An approach to mitigate black hole attacks on vehicular wireless networks. In: 2017 IEEE 85th vehicular technology conference (VTC Spring) (pp. 1–7). IEEE, June 2017

14. Alsulaim NA, Alolaqi RA, Alhumaidan RY (2020) Proposed solutions to detect and prevent DoS attacks on VANETs system. In: 2020 3rd international conference on computer applications & information security (ICCAIS) (pp. 1–6). IEEE, Mar 2020

15. Wararkar P, Dorle SS (2016) Transportation security through inter vehicular ad-hoc networks (VANETs) handovers using RF trans receiver. In 2016 IEEE Students' conference on electrical, electronics and computer science (SCEECS) (pp. 1–6). IEEE, Mar 2016

16. Malhi AK, Batra S, Pannu HS (2020) Security of vehicular ad-hoc networks: a comprehensive survey. Comput Secur 89:101664

17. Zhou M, Han L, Lu H, Fu C (2020) Distributed collaborative intrusion detection system for vehicular Ad Hoc networks based on invariant. Comput Netw 172:107174

18. Manivannan D, Moni SS, Zeadally S (2020) Secure authentication and privacy-preserving techniques in Vehicular Ad-hoc NETworks (VANETs). Veh Commun 25:100247

19. Chuprov S, Viksnin I, Kim I, Reznikand L, Khokhlov I (2020) Reputation and trust models with data quality metrics for improving autonomous vehicles traffic security and safety. In: 2020 IEEE systems security symposium (SSS) (pp. 1–8). IEEE, July 2020

20. Agrawal R, Chatterjee JM, Kumar A, Rathore PS (2020) Blockchain technology and the Internet of things: challenges and applications in bitcoin and security. Apple Academic Press. https://books.google.co.in/books?id=FCoMEAAAQBAJ

21. Abassi R, Douss ABC, Sauveron D (2020) TSME: a trust-based security scheme for message exchange in vehicular Ad hoc networks. HCIS 10(1):1–19

22. Jenefa J, Anita EM (2016) A signature-based secure authentication framework for vehicular ad hoc networks. Int J Comput Inf Eng 10(2):403–408

23. Joshi Y, Joshi A, Tayade N, Shinde P, Rokade SM (2016) IoT based smart traffic density alarming indicator. Int J Adv Comput Sci Appl 3(10):1086–1089

24. Mármol FG, Pérez GM (2012) TRIP, a trust and reputation infrastructure-based proposal for vehicular ad hoc networks. J Netw Comput Appl 35(3):934–941

25. Gurung S, Lin D, Squicciarini A, Bertino E (2013, June) Information-oriented trustworthiness evaluation in vehicular ad-hoc networks. In: International conference on network and system security. Springer, Berlin, Heidelberg, pp 94–108

26. Adimoolam M, John A, Balamurugan NM, Ananth KT (2021) Green ICT communication, networking and data processing. In: Balusamy B, Chilamkurti N, Kadry S (eds) Green computing in smart cities: simulation and techniques, Green energy and technology. Springer, Cham. https://doi.org/10.1007/978-3-030-48141-4_6

27. Li W, Song H (2015) ART: an attack-resistant trust management scheme for securing vehicular ad hoc networks. IEEE Trans Intell Transp Syst 17(4):960–969

28. Ghori MR, Zamli KZ, Quosthoni N, Hisyam M, Montaser M (2018) Vehicular ad-hoc network (VANET): review. In: Proceedings of the 2018 IEEE international conference on innovative research and development (ICIRD), Bangkok, Thailand, 11–12 May 2018

29. Hasrouny, H., Bassil, C., Samhat, A. E., & Laouiti, A. (2015). Group-based authentication in V2V communications. In 2015 fifth international conference on digital information and communication technology and its applications (DICTAP) (pp. 173–177). IEEE, Apr 2015

30. Florian M, Finster S, Baumgart I (2014) Privacy-preserving cooperative route planning. IEEE Internet Things J 1(6):590–599

31. Li J, Lu H, Guizani M (2014) ACPN: a novel authentication framework with conditional privacy-preservation and non-repudiation for VANETs. IEEE Trans Parallel Distrib Syst 26(4):938–948

32. Mustafa AS, Hamdi MM, Mahdi HF, Abood MS (2020) VANET: towards security issues review. In: 2020 IEEE 5th international symposium on telecommunication technologies (ISTT) (pp. 151–156). IEEE, Nov 2020

33. Bhatia TK, Ramachandran RK, Doss R, Pan L (2020) A review of simulators used for VANETs: the case-study of vehicular mobility generators. In: 2020 7th international conference on signal processing and integrated networks (SPIN) (pp. 234–239). IEEE, Feb 2020

34. Pavithra, T., & Nagabhushana, B. S. (2020, July). A survey on security in VANETs. In 2020 second international conference on inventive research in computing applications (ICIRCA) (pp. 881-889). IEEE
35. Rajmohan, R., T. Ananth Kumar, M. Pavithra, and S. G. Sandhya. Blockchain: Next-generation technology for industry 4.0. In Blockchain Technology, pp. 177-198. CRC Press, 2020.
36. Kumar, K. Suresh, T. Ananth Kumar, A. S. Radhamani, and S. Sundaresan. Blockchain Technology: An Insight into Architecture, Use Cases, and Its Application with Industrial IoT and Big Data. In Blockchain Technology, pp. 23-42. CRC Press, 2020.

Blockchain Technology in Healthcare

Anita Khosla, Shruti Vashist, and Geeta Nijhawan

1 Introduction

The idea of the blockchain itself is basic; it is a system of blocks that is ceaselessly developing and dynamic in nature, adjusting to the demands of particular businesses and their quirks. The blockchain depends on the distributed ledgers containing data or certainties. As it were, a block keeps a record of all interactions. At a point when the new block is completed, it is added to the series of recently made blocks. In this manner, a tremendous arrangement of blocks with information is made, where data are saved and can be fetched whenever required.

In the monetary circle, banks utilize the blockchain as a decentralized record for progressively secure and quicker exchanges. In the data innovation industry, blockchain is utilized as a reason for digital storages, software programs, data platforms, and so forth.

Healthcare and medicine are additionally experiencing changes because of the blockchain. The established innovation utilized in healthcare is brought together. In any case, the circle is currently encountering advantages of the decentralized system

A. Khosla (✉)
EEE Department, Manav Rachna International Institute of Research and Studies,
Faridabad, Haryana, India
e-mail: anitakhosla.fet@mriu.edu.in

S. Vashist
ECE Department, Manav Rachna University, Faridabad, Haryana, India
e-mail: shruti.fet@mriu.edu.in

G. Nijhawan
ECE Department, Manav Rachna International Institute of Research and Studies EEE
Department, Faridabad, Haryana, India
e-mail: geeta.fet@mriu.edu.in

© Springer Nature Switzerland AG 2022
P. Raj et al. (eds.), *Blockchain, Artificial Intelligence, and the Internet of Things*,
EAI/Springer Innovations in Communication and Computing,
https://doi.org/10.1007/978-3-030-77637-4_9

Fig. 1 Healthcare
challenges

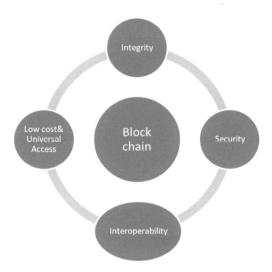

too. For instance, putting away different medication-related information has turned
out to be progressively secure because of blockchain [1].

The blockchain is additionally as of now being used for document moves in
clinical fundamentals, patient's information and pharmaceutical history storage,
and creation and upkeep of the blockchain coordinating with patients, doctors, and
payers with further assurance of CMN by methods for smart contracts, in this manner keeping away from intermediation.

Blockchain furnishes us with an unfathomable opening to beat different difficulties present in the healthcare industry today, like interoperability, security, honesty,
discernibility, and widespread access.

Blockchain addresses the current challenge for syncing patient data in health
information system (HIS) among multiple unrelated data; while doing this, a distributed framework is adopted for ensuring patient data security and privacy to manage patient uniqueness [1]. Few healthcare challenges addressed by blockchain are
shown in Fig. 1.

1.1 Interoperability

As of today, if interoperability of healthcare data is taken into account, then healthcare organizations are at different maturity levels. The use of FHIR is explored by
some of the organizations, while the CDA standard data exchange is used by others,
and data are also being shared by using HL7 2.X standard. That different organizations are using various data standards is another challenge that hampers quality
score interoperability. Blockchain may be used to access data through APIs to overcome this challenge. While transferring the data using APIs, data format gets

standardized and then the same is transmitted to communicate different HL7 versions irrespective of the capabilities of EHRs.

1.2 Security

In the healthcare industry, security is chosen by different angles dependent on the unwavering quality of information kept up in an association. For many healthcare organizations, tampering and breaching of security of healthcare data are becoming a growing concern. The advantage of blockchain is that data present in it are difficult to tinker because it needs confirmation from various nodes included in the chain. The data in blockchain may be encrypted using a private key of the sender [2]. Therefore, blockchain has embedded security features and data can be decrypted only by who is receiving using key sent by the sender.

1.3 Integrity

The data integrity is ensured and helped by blockchain's scattered ledger and undeniable transactions, whereas the security of data across the network is improved by encrypting the data. The advantage of implementing blockchain technology is that outdated data of patients available with different organizations are updated and replaced by the latest information of patients. Only the patient who can control the data will be the custodial owner.

1.4 Low Cost

Support of a specific healthcare information system includes distinctive activities that however are not restricted to performing reinforcement stockpiling gadget fields. Information is disseminated over the system, and in the event that there is a failure at any single point, then it prompts a normal backup instrument [3]. Notwithstanding, this single adaptation of information is refreshed on every hub of the blockchain. It decreases the measure of exchange between every data framework, subsequently helping the weight on the healthcare system.

1.5   Universal Access

The main challenge in healthcare organizations is administering access to patient data. It is ensured by blockchain that at every node data required are available for the use of authorized entities as they are given access rights via smart contracts.

2   Blockchain in Healthcare Today

2.1   What Has Blockchain Changed in Healthcare?

The pace at which innovation is advancing requires a need to bring capable healthcare confirmation frameworks, wearable gadgets, medical examination systems based on artificial intelligence, and more innovation to meet the growth. The working of medical clinics must incorporate cryptography. The necessity of growth escalates to the top level when it is related to healthcare. Today's foremost requirement is the quality of healthcare services. Furthermore, the healthcare system is based on a patient-centric approach directed toward affordable treatment and pertinent healthcare facilities [2].

In order to focus on quality healthcare services, one needs to ensure that management of healthcare of patients is given top priority. But policies that are drafted at the central level make processes even more tiresome and protracted. So, in order to keep processes like this integral and tranquil, effectual patient care is not realistic in many cases. The presence of middlemen is another factor to degrade it further.

In the healthcare systems, data of all the patients and information including the critical ones also remain speckled among various systems and departments. Because of this, sometimes-vital data are not accessible when required the most. Numerous players in the system have not planned for smooth process management, so the existing healthcare ecosystem cannot be deemed complete. The major difficulty in the way of quality healthcare services is the split between provider and customer, which in this case is patients. What makes it even worse is the presence of middlemen in the supply chain.

Many healthcare facilities are using obsolete systems even today for keeping data and records of patients, which also includes the functionality of maintaining local records.

It becomes very tedious and time consuming for the doctors to diagnose the problem and likewise difficult for patients as well. Due to all these factors, it became more desirable to adopt the approach of patient-oriented business. These data are insufficient even for the exchange of information and need some key changes.

Mishandling of present data by healthcare organizations is another reason, which is keeping them away from imparting proper patient care and high-quality services to obtain better health. Otherwise, these organizations are quite efficient if seen in the context of economy but still not able to meet the expectations of patients. Some observations on these facts are the following:

- Around half of the clinical treatments globally are not recorded.
- Also, nearly half of the data records provided by healthcare providers have errors or have incorrect information.
- Gaps in organizations regarding healthcare data are assumed to be high at present, which is anticipated to rise further in the future.

These are not the only issues in the current healthcare sector systems. The problems get increased with the passage of time. So, it becomes pertinent to have a technically advanced system. For example, the problem of drug counterfeiting leads to huge losses. It could be reduced considerably by placing a system with precise tracking features in the supply chain.

Blockchain innovation is these days recognized for its revolutionary effect on healthcare services compelling its digitalization and change. It is occurring worldwide with the most extraordinary tasks of blockchain innovation in healthcare services to be talked about.

2.2   The Most Innovative Blockchain Projects in Healthcare

Among the most inventive endeavors including blockchain, there are many blockchain systems in healthcare; the most obvious of them are as mentioned below:

- MedRec saves the computerized family history of medicinal accounts using methods included in blockchain.
- The MediLedger utilizes blockchain with respect to doctor-suggested medicine. It enables clients to make an interoperable framework for recognizable proof and following explicit physician endorsed medications. This meets the law and operational needs of the business [5].
- Connecting Care is one more blockchain-based platform for suppliers together from various clinical associations to get similar information for patients shared among them. This project is reliable and can be adopted by healthcare services industry, which can further help the advancement of the marketplace in healthcare data.
- Robomed Network is a medical network based on blockchain intended to give the best medicinal consideration. It will associate patients and healthcare service specialist organizations with methods for smart contracts. It is projected to support the move of healthcare services to outcome-based consideration (when a patient pays for results as opposed to paying for methods). Robomed, as of now, is having around 9000 patients and 30,500 administrations.

In the near future, to utilize the technology in healthcare, blockchain will be among front runners alongside the Internet of things and AI to prompt new and progressively compelling digital processes. It will bring revolution in healthcare.

Truth be told, it began to do as such. Due to the advancement and pervasiveness of blockchain innovation, it will be easier to persuade healthcare providers to convey the technology for the savvy putting away of information, distinctive exchanges, and temperature and complex programming-driven medicinal appliances, which are under consideration and investigation.

2.3   The Need for Blockchain in Healthcare

One area where blockchain has definitely unbelievable potential is healthcare [4]. To change healthcare, the organization of data should be given more focus so that it can benefit in relating notable systems and play a major role in the accuracy of EHRs. Progress of blockchain can also be helpful to suggest medicines and SCM (supply chain management) and any hazard statistics administration. More healthcare regions that can be benefitted using blockchain advancement are benefactor authorizations, medicinal charging, constricting, medical record discussion, and medical preliminaries against the piracy of treatments.

Healthcare facility administrations take all efforts to allow a methodology that is patient-driven. Healthcare systems based on blockchain could get better security and dependability of patient's statistics as patients might have a direction for healthcare histories. Such classifications can assist in solidifying understanding information and also help in providing medicinal histories among various healthcare establishments. Keeping the medicinal data of patients is crucial in healthcare systems. Such information is exceptionally flimsy and also applied objective for cyberattacks. So this makes it a basic need to verify complete information. Another perspective is the authority of information ideally directed by the patient. In such a manner, sharing and dealing with patients' medicinal service information is another usage case that can be benefitted by forefront present-day developments [8].

Blockchain innovation is enthusiastic against outbreaks and disappointments and also offers different ways for control. Along these lines, blockchain gives a not-too-bad framework to healthcare information. For individual medical data, the most attractive thing is a secluded blockchain. The choice model by Würst and Gervais conveys that a blockchain is the best option when numerous congregations not sharing with others require to associate for exchanging common data. They do not wish to contain a TTP (trusted third-party) [5]. The model exhibits some variables, which are to be viewed in detail if a specific condition entails blockchain.

Concerning storing, numerous features are to be measured [5]:

- Is it necessary to store data?
- Is there a need for multiple write access?
- Is a "TTP" available and is there someone who is constantly connected is employed? First thing, we need to choose the prerequisite for data taking care of data set. Subsequently, it ought to be settled if there is a necessity for making access for various social congregations. If simply one author exists, no blockchain

is required and elective plans might be thoroughly examined (a database). One more thing to be noticed is that better execution is offered by conventional databases than a blockchain. For an event, if a "TTP" is available constantly online, and one could rely on it, then blockchain is not required at that point. The "Würst and Gervais" choice model furthermore sorts out the type of blockchain needed (e.g., open consent less, open permissioned, or private). In the event that in an occasion that the writers are obscure, the key choice would be an open approval barring blockchain. If the TTP is detached, it could work correspondingly to a confirmation master and the included assemblies don't generally trust in each other; a permissioned blockchain could be used. Regardless, if all of the gatherings ordinarily trust each other, a data set with shared access could be used as opposed to a blockchain. Of course, if writers are known and can be believed, the choice falls between an open permissioned and private blockchain. The first is for the circumstance where one necessitates open evident nature and the second is for the circumstance when it isn't needed. The current helpful data system, by and large, depends upon trusted outcasts. In a couple of cases, these can't be totally trusted [6]. The blockchain, which relies upon accord and doesn't need a central subject matter expert, is a possible response for this issue.

3 How Blockchain Is Changing the Future of Healthcare

In spite of the fact that joining of blockchain and healthcare is a tedious process, the capability of consolidating them can't be overlooked and it has just begun. In spite of the fact that blockchain is being embraced by healthcare services industry at an extremely moderate pace, in the coming years it is without a doubt that healthcare IT is set up for the inescapable eventual fate of the business [9].

As the technology is growing rapidly, soon the IoT devices can collect the basic physiological parameters of the patient like pulse rate, blood pressure, body temperature, and so on and that too in real time. This gets appended in the data of the patient, that is, electronic health records. IoT and blockchain go hand in hand. IOT can help the blockchain grow, and indeed, blockchain technology can help improve the prospects of growing IoT world. The detailed in-store information is sent continuously [7]. The doctor will be able to monitor the data and will be able to give the correct diagnosis based on the exact details of the patient parameters. Eventually, he can start the line of treatment. While exorbitant presently, blockchain can be the following innovative light to upset what's to come. For pharma, it could mean getting genuinely necessary trust between parties with interest in medication inventory network honesty pharma organizations, producers, exchanging accomplices, drug stores, and patients with the additional advantage of re-upholding consistency with administrative bodies [10].

Blockchain innovation is ready to address the significant trust that gives forestall sharing of well-being data. Due to assent and protection issues, one's well-being data are as of now segregated. Blockchain is the central innovation that is giving the structure to empower growth for the utilization of electronic health records (EHRs) using transformational applications.

Earlier, the data of patients were moved by paper, which currently shifts immediately over "TCP/IP" (transmission control protocol/Internet protocol) and blockchain advancements. TCP/IP is the protocol architecture of the Internet, the system on which EHRs are based, just as email. Blockchain is the "future" correspondence convention for system administration. This structure shall institutionalize EHRs in the upcoming years and in the long run outcome in the making of the national well-being data framework imagined. Blockchain technology can possibly change applications of healthcare and the interoperability of EHRs in the near future. Blockchain shall organize well-being data frameworks to trade and make utilization of EHRs found anyplace in the not-so-distant future, anywhere on the planet.

Nowadays, blockchain does not do much in healthcare. But in the coming times, it will. Blockchain technology will soon revolutionize healthcare.

1. It is difficult to administer health records, which requires a high operating cost, including the risk of human mistakes. As per a national Mayo Clinic, the main thing which most of the doctors would like to reform is the health record process [7]. The purpose behind this oppressiveness is on the grounds that well-being data are not situated in a solitary database yet rather disseminated among various on-screen characters who claim and trade the information for each and every patient(https://www.usfhealthonline.com/resources/healthcare/does-blockchain-have-a-future-in-healthcare/).

Record frameworks are divided and are also helpless since essential PC security conventions that verify persistent information are frequently deficient. One of the examples of powerlessness is that it is not required by the national government all the inclusive encryption of well-being records yet rather arranges suppliers for utilization of a dimension of security that is sensible yet appropriate [8].

On the other hand, clinics also give information of cyber-attacks on the records stored, which appear in the introduction of medicinal personal or money-related data, for example, malware that takes information. Not considering the information encryption of ransomware, a program hides sufferer details until suppliers give a payoff [9].

Blockchain technology can be used to improve various configurations as explained:

3.1 Connectivity

First, unlike the wellness data systems implemented today, which were intended for a closed case without thinking of larger applications, blockchain technology is intended for availability and circulation in an open system. By ensuring that data are understood within a mutual framework, blockchain maintains a vital separation of risks from securing prosperity records in a region. This is considered a trusted third party, for example, a clinic containing a patient's medical history or a welfare protection plan containing a patient's requests for human services. If that unknown bears a catastrophic information problem, or even a minor one, such as contamination of information by a programmer, it could be embarrassing or even difficult to prove or deny a patient's restoration story [13].

3.2 Ledger Technology

Healthcare suffers from inefficient operations, but blockchain and shared ledger technologies can ease the pain and take the industry to a whole new level. Exchange issues maintain a strategic distance from a conveyed record since when some information in a block is modified or adjusted, then it winds up unacceptable, and the ensuing arrangement of connected information in blocks ends up untrue. It is on the grounds that the data of every datum block are utilized in an arithmetic capacity to produce the connection in a chain to the following block. In that capacity, modifications are difficult to develop. Instead of having one focal overseer of well-being records, which goes about like a watchman to persistent information—a rundown of computerized exchanges—one shared encoded record is there, which is spread over a system of matched, duplicated databases, which are straightforward to anybody with admission. Remarkable security benefits are provided to the system using it. To hack a single block from the given chain is incomprehensible by not hacking every block in the sequence of chain provided (in this way, the term blockchain). Thus, this makes blockchain speaking to the doctors and clinic frameworks where secure access is needed to a patient's health history. The well-being records might be exclusive and at the same time originated from better places, yet the appropriated record itself is institutionalized [21].

3.3 Precision with Pseudonymity

Blockchain innovation can also improve transparency with pseudonymity. Pseudonymity (for example, date of birth and postal district) happens when a patient is recognized by an option that is different from their genuine name; it can be applied to any record of well-being exchange of patients have, which shields their uniqueness to informed to other groups. Diverse dimensions of pseudonymity are present, and instances of pseudonymity are visible all over EHRs [14]. A well-being record regarding this structure is appropriate for broad investigation and handling.

3.4   Secure Data Encryption

The next normal for blockchain innovation is its security. The information in a block, or the EHR, could be put away in an encoded structure utilizing open key cryptography and may be opened utilizing a secret phrase which patients, exchange proprietors have. This secret phrase would make it unthinkable to access unapproved health data. Now the subject of what occurs in a crisis when a patient might be weakened as well as powerless to give the secret phrase is regularly presented. A few alternatives exist, including crisis outsider backers with access to a secret phrase, which may be utilized by follow-up through a blockchain expert and will stand out from all gadgets (e.g., sensors, wearables, and patient health applications).

3.5   Security

The EHR security is developing intense as malware and ransomware multiply. The irritating truth is that even the genuine condition regarding cybersecurity hazards in electronic health systems is not reported properly by requesting many times. Though the ability of services looking after healthcare systems to battle cybercrimes is controlled with TCP/IP innovation, this safety issue could be inclined to blockchain innovation [15].

Blockchain innovation, similar to TCP/IP, is a primary innovation that can empower transformative change and advancement in human services [24].

3.6   Unified Clinical Systems

In general, there are four parameters of block data—connectivity, register technology, precision with a pseudonym, and data security— and it is important to ensure the security of the EU and guarantee the quality of information.

Much the same as how the Internet changed the manner in which health records are shared between clinics and doctors, blockchain is an open-source development to reform health record exchanges among healthcare systems and patients. Blockchain innovation would make regular information that is

- Exact, implying that the correct information will be practical and explicit.
- Absolute; as such, health record of a patient will have all the required information.
- Predictable, which means information of the patient can be used crosswise over various sources, various suppliers, and also over the different pieces of a system pertaining to healthcare.
- In time for actual, information-motivated choices.
- Remarkable and legitimate.

As blockchain, bound together clinical frameworks can be made with more prominent two-way precision and lesser working costs [11]. Doctors who deal with their health records accurately and appreciate their intrinsic value would be having a preferred standpoint in the marketplace.

3.7 Transformational Law for Keeping Health Data

Blockchain innovation is having the most troublesome effect on data built on well-being ever since the development of the digitization. Far-reaching well-being data rules are projected to suit innovative changes that happened in the course of recent time to be precise since the Internet unrest initially started in 1965 [12]. In the future, whether blockchain technology is embraced or not, but the global technology revolution warrants consideration regarding blockchain solutions [26].

4 Benefits of Blockchain in Healthcare

The healthcare industry has excess data to handle. It consists of clinical records, patient history, multifaceted billing information, health checks, and so on, which make it intricate to manage not beyond the healthcare industry methodically [16]. According to research, people see a large number of different doctors in their lifespan. It becomes extremely difficult and impractical to assemble and administer the whole information individually [17].

Digitization of health records has assisted in converting the arduous manual information into digital data. This expertise has supported many medical practitioners to handle the data of their patients in a better way. The digitized data of the patient is termed as the EHR or the electronic health record [20]. These electronic health records aren't generally shared between systems. Initially, there was not a single reserve on hand where all individual patient information could be available. Currently, the doctors are able to avail any of the three mentioned techniques to access information.

- *Push*: Medical information of the patient is transferred from one doctor to the other [27].
- *Pull*: One source can request the information from the other.
- *View*: Provider with rights can analyze the data inside other provider's record.

Finally, the time has come that an unsullied approach needs to be thought about to improvise the information management in the healthcare industry. There is a need to develop a protected site where the data of the patient along with his/her complete medical history and pertinent information could be stored and saved methodically for further use and investigation. Blockchain is a lone expertise that can deliver the key to cater to the need of the hour. It can be thought of as a future of data management in the healthcare industry [20]. Let's thoroughly examine the blockchain concept and its benefits in the healthcare industry.

Various cryptographic techniques have been applied in the development of the blockchain. Due to this cryptography, there will be no requirement of the central manager. All the service providers have total control of all their information and communication. It has been observed that the healthcare deals with classified individual information about every patient and it requires swift access to this information; blockchain reorganizes these desired data and facilitates their allocation in a secure way. Thus, the blockchain offers security, scalability, and data privacy [24].

One of the major highlights that make the blockchain novel and innovative in the healthcare sector is the absence of a central officer. If the information pertaining to the individual patient is stored in a certain system, any undefined person having the right of accessing the particular system can tamper the statistics. Every minute detail related to the patient data, right from registry of the disease, laboratory results, and line of treatment, could be attained using blockchain, which includes patient history, outpatient care, wearable data, and supporting the providers in better ways for delivering care [24].

5 Patient Database for Healthcare Sector

Records usually are uneven or reproduced while dealing in healthcare. Similarly, EHRs take various schemes in each field having diverse methods to enter and then manipulating records [26]. Using blockchain, the complete record is diced to a ledger. The user will search for a particular desired location. It could have several locations and number of solutions associated with it, but all of them tell about the same patient credentials [6, 27].

- *Claim Arbitration*

As it has been observed that the blockchain works on a validation-based exchange, the claims can be automatically verified where the network agrees upon the way a contract is executed. There are very few errors or frauds since there is no central authority.

- *Interoperability*

  - The blockchain's key feature is interoperability, and it can be comprehended by using sophisticated application programming interfaces (APIs). This makes electronic health record interoperable and information storing becomes a secure procedure. As the blockchain network is common with approved benefactors in a safe or homogenous mode, it eliminates the charge and the herculean task related to data compilation [19].
  - Blockchain can also alter the revenue cycle and drug supply management, along with the medical trials, and thus prevents thefts.

In one or the other way, the healthcare is being taken by blockchain as the storm over the past few years, so there has been a momentous growth in the investments for blockchain. Thus, with this possibility, no wonder that blockchain has brought a revolution in digitization. Perhaps one day, it will change the big data scenario [25].

In healthcare, humongous data are spread among different schemes, and at times, it might be unavailable when required. The present healthcare communication in today's scenario is ill planned, and there has been difficulty in handling the exchange of the information. There is a need to change the way we view healthcare and data together.

The drastic revolution in the use of blockchain technology in the healthcare industry has increased the number of applications; a few of them are listed below [24]:

- *Community Health Service*: The use of blockchain technology helps the authorized bodies to create and share a stream of patient information that is de-identified. This is mainly the process that is adopted to maintain data privacy [29]. This also helps the authorities to identify the threats so that timely necessary action could be planned to control the problem well in time.
- *Data Security*: Blockchain provides a smart technology and prevents unauthorized individuals from accessing the information.
- *Consent Management*: It is a system that sets certain policies that allow the patients to determine what health information the providers can access with their permission. Patients can exclusively permit any individual to access their medical information.
- *Simplified Claim Processing*: The blockchain technology can simplify the intricate billing process in the health sector by eliminating the series of validations.
- *Patient-Generated Data*: Using the blockchain technology, the patients will be able to easily upload and store their restructured medical information without messing up any previous records. The data uploaded will be secure.

5.1 Blockchain Opportunities for Health Data

The implementation of blockchain information and skills can resolve many problems that the pharmacy, health, and medical industries are presently experiencing. Some of these problems related to health data can be in terms of protection, interoperability, and ease of access in context to the medical investigation, experimental trials, medical supply chain, and drug reliability [18, 19]. The expertise has potential in terms of managing therapeutic data, preventing breaches, enhancing interoperability, reorganization of the treatments, tracking drugs and prescriptions, and taking follow-ups of the supply chains and healthcare devices. Blockchain technologies in the sector of healthcare have great potentials to explore. The growth and development in the medical and health sector using the technology of blockchain technology is here to stay.

The utilization of blockchain technology in healthcare can bring revolution in the field of data management. The technology can make the exchange of medical data more efficient, secure, and transparent across the healthcare sector. It also helps in maintaining and updating the files.

This removes the unwanted clerical costs and thus allows appropriate data utilization of physical condition. Moreover, the use of the technology can, in a way, reduce the need for various intermediaries to overlook the exchange of vital health data [30].

As mentioned before, the medical data of a single patient can be bifurcated into many facilities like caregivers and insurance providers. It suggests that the complete medical history of the patients is generally unclear and not complete. The medical record that is stored digitally in a blockchain can help the facility provider of the given sector to assemble desired user information of the health data together. Thus, the complete digital information of the medical patient, that is, their medical prescriptions, line of treatment, facilities used, and any other relevant information, would always be prevalent for the concerned authorities.

This could, in a way, ease the task of the healthcare worker whose job is to offer professional and appropriate treatment to patients. Implementation of this technology in the healthcare sector would help the treatment providers to have a complete picture of any given patient's medical records and information. The most important and vital aspect of this technology in the domain of the health sector is that all the past data and records of all the patients are unalterable and any alterations and changes in the record or information will be visible very clearly.

Management and organization of the medical data and information are one of the most valued advantages to blockchain technology in the sector of healthcare. The medical industry has been facing many issues like interoperability, insecurity, data scam, and even loss of data during adversity. All of these could be eliminated by implementing blockchain technology.

1. The main highlight of this technology is disaster recovery. The health data are never secured and stored in one location, and thus any major failure of one location will not corrupt the whole information. Due to any of the natural calamity or disasters like earthquake, fire, flood, and asteroid strike, the data would still be safe (http://blog.appliedinformaticsinc.com/blockchain-transparency-integrity-and-clarity-to-the-healthcare-system).

To be able to provide accurate and proper diagnosis and healthcare treatment, it has to be ensured that the health data are appropriate and correctly managed and stored. Right and secured access to this medical information will again reassure proper health diagnostics. It can further be reassured by the information that once any data are uploaded using this technology, any change or alteration to them become nearly impossible. The patient information can be stored in a blockchain in the form of documents, wearables, handheld devices, EMRs, and so on.

5.2 Incorporation of Blockchain Technology in Healthcare Sector

Blockchain is very effective in creating a solo patient entry, that is, the electronic health records. Access to the system is easy any time anywhere, and so the patient should not worry about the organization of the information. With this revolution, any practitioner will be able to review the medical history and recommend the line of treatment, that too with the agreement of the patient. All the data which is added is logged as an all-purpose document irrespective of being or through any company.

It has been observed that often in the patient's history, the private doctor's appointment is never mentioned. This is necessary as it presents a clear picture of the patient's illnesses. With the rapid growth in technology, this information is always appended in real time, and the operation is carried out instantaneously. With this, the doctor will be able to track basic health parameters related to the patient like blood group, chronic diseases, allergic reactions, and so on from his/her previous history. This will indeed reduce the burden of repeated tests and thus save time and improve the speed of treatment.

Drug Tracking

The medical sector has been facing the problem of fake products. The implementation of blockchain can solve this issue. Due to the digitization of data, each deal between the firm and the vendor can be put on record in a network. It allows the system to authenticate the medicines and will in no time shares the details of the maker and all the previous transactions done.

Consistency

The use of blockchain technology can bring some amount of standardization, uniformity, and reliability due to standard protocols and electronic health records. The consistency of data allows maintaining a single global standard.

6 Implementation of Blockchain Technology

Blockchain technology possesses the authority to bring an enormous revolution in the medical management sector. This has revolutionized the medical management in the life of a common person. Everyone shall be accountable for managing individual data and records and thus can manage their own information.

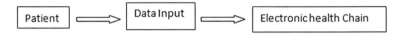

Fig. 2 Implementation of blockchain

The advancement in the expertise in the field of blockchain caters to the efficacious improvement in the quality in the medical maintenance keeping the capitals to a minimum. The electronic data are maintained, as shown in Fig. 2.

The interventions occur at many stages of verification and can be eradicated through blockchain. Even though the technology is in its initial stage, it is popularly acknowledged by the public in the medical sector and it can be implemented [9].

6.1 Distributed Network

Blockchain opens new prospects in the healthcare sector. It stores individual data of patients, which are in the form of electronic health records. This technology is implemented using distributed networks. The data and the desired information are not secured in one location. The information is duplicated and stored in encrypted form for each user.

This rejects the likelihood of distributed denial of service attacks (DDoS). The hackers who can eavesdrop cannot demolish or replace data. Even if one among all the systems is working and the rest all fail, the transaction would be carried out successfully and the chain will always remain available. Backup of all the transactions is done systematically with high security. A high level of encryption based on secret private keys eliminates the possibility of eavesdropping on any information.

6.2 Step-by-Step Recording

Step-by-step is one of the significant methods of documenting the data. In the string of data, one can augment a new portion but cannot remove the existing one. The algorithm associated with the blockchain operation permits the hacker to alter the existing data and record the existing procedure; he will require the authorization of other computers (users) of the system. This occurs in a mechanical and instinctive mode. If any constituent of the chain does not match the same constituent on another computer, the record operation is canceled. Thus, the whole chain must be complete. No one can alter the previously recorded information.

Due to this, the patient is always sure that his information is always secure and cannot be altered by anyone. Along with this security aspect, the other thing associated is the update and entry of the data at the doctor's end. The data must be filled with utmost responsibility. Any mistake or error at the doctor's level can lead to an unlikeable situation.

6.3 Global and Worldwide Access of Information

The blockchain technology has worldwide accessibility. The patient can access the treating doctor; similarly, the doctor can get information about the patient being treated and from anywhere and anytime in the world using permissioned blockchains and permissionless blockchains, as shown in Fig. 3.

6.4 Permissioned Blockchains

As the name suggests, these types of blockchains allow the information to be exchanged in real time among the various contributors of the system, that too on a permissioned basis. It is a system in which all the individuals work in a closed environment, and all of them are able to interact with each other and can approach the network. The transaction of data between the individuals can be done within an organization or institution in a secured way. The exchange of information once done will be administered through agreement, and it will be preserved as permanent information or data, and it gets appended as a new addition to the prevailing blockchain.

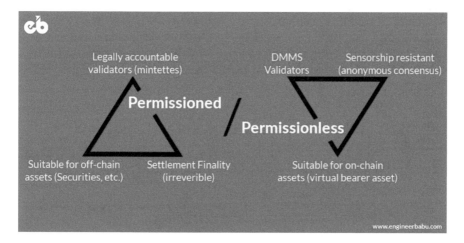

Fig. 3 Comparing permissioned and permissionless blockchains [28]

6.5 Permissionless Blockchains

As compared to the permissioned blockchain, the permissionless blockchain permits access to the user and allows the user to generate their own address and start with the interaction instantly in a given network. An example of this kind of permissionless system is the Internet at today's time. The Internet permits any individual to design and create their own site. Just by merely creating an address on any site, the permissionless blockchain allows the interaction with other participants on the same network [4].

Among the above, the blockchains could be effectally cast off in healthcare to take the correct choices within the healthcare network. There is a lot of potential growth associated with the blockchain in healthcare, and this could be investigated further. There are a lot many potential sectors that could be explored further.

Examples of Implementation of Blockchain Technology

Drug Traceability

According to the survey given by the World Health Organization (WHO), 20% of the drugs in emerging countries are not original, which leads to a major economic and financial loss every year [27]. The tracking of a drug is shown in Fig. 4. This is the major cause of death in many countries due to the wrong intake of medicine. To tackle this issue, the security aspect of this blockchain technology can be utilized in drug traceability and thus security. Every new deal that is added to a block will be undisputable and will be time stamped. Due to this enhanced feature, it becomes simple to ensure a product and guarantee that the data inside the chunk cannot be altered to add to the security aspect of the system and to certify the traceability and legitimacy of the medicines. Medicinal organizations that register any specific medicine drug on the blockchain need to be reliable and dependable. The systems are such designed that the reins of all the private blockchains are in the control of a central authority. If the drug manufacturers are able to provide an authorized certificate that the drugs developed by them are authentic, only then they are given access to the specific drug blockchain.

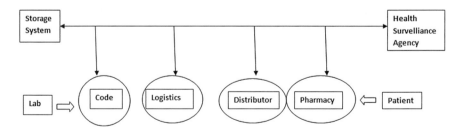

Fig. 4 Drug traceability from storage to agency

All the drug laboratories can hold the records for various drugs. The dealers or the vendors have the right to authenticate a deal. All the information related to drugs is interlinked within all the blocks. The movement of drugs along with various dealers and retailers have to move along the supply chain among different articles and it can be easily stalked. This makes the complete system very transparent and thus has information about the movement of fake drugs [22].

Medical Prosecutions and Data Security

In order to certify and investigate the efficiency of any system, medicinal trials are conducted and then developed for healing an explicit ailment. On the basis of trials and tests, the drug is developed and verified, and depending on the success of the trial and testing, it can be implemented on a larger scale. It is clarified from above that in order to run successful trials for the actual development of drugs large amount of data is required. The investigators continuously develop and emphasize the information and regularly conduct tests under different settings to generate data, figures, and efficiency. Based on these tests and trials, the data can be further analyzed.

In the cutthroat competition today, most of the researchers are developing, hiding, and modifying the collected data. Most of the pharmacological companies today are showing keen attention in recording the results that can assure certain benefits for their firms.

In a directive to make the medical hearings fairer and further clear, investigators can adopt blockchain technology, which helps develop a flawless system with a high level of security that will be able to maintain documents in an unprejudiced way for medical trials. The maintenance of data is an arduous exercise. The current data need to be appended with the existing data, which could be guidelines, protocols, research plans and proposals, and so on. But the basic requirement is that they all should be time stamped. This means that all the above-mentioned documents should have a proper verification and authentication at the time of creation so that time stamping could be done. For premeditated data, it is explicitly vital to keep the information time stamped to generate a proof that displays that the pact was there even before the trial started.

Blockchain technology enhances the reliability of scientific and medical trials and analysis. The data can be stored as smart contracts on the blockchain, substituting the digitized impressions. This collection of documents minimizes the cost for review of documents, files, thefts, and lost-found data. Along with the above, the blockchain also keeps supervision and management of the supply chain and the liability of drug tracking.

Implementation of this blockchain to these vast data related to the health sector solves many concerns. The initial individual patient data management system had created a clutter. A blockchain system replacing the existing manual data management would create systemic information with proper time stamping and security. The maximum time for which a third party can access the information is also controlled by the individual patients. The efficiency, reliability, accessibility, and security of the system improve as the manual data are taken over by digitized data [23].

7   Challenges of Blockchain in HealthCare

The primary hold on using the technology could be pretty inspiring; therefore, it is sensible to interact with experts for a proficient discussion on the future of blockchain in healthcare. Blockchain's future in healthcare is subject to how responsible healthcare organizations are to develop the required technical infrastructure. No doubt, the development of blockchain is a revolution, but at the same time, the system involves cost and there may be some issues concerning its combination with the prevailing technology.

Because of the tremendous outbreak of communicable and noncommunicable diseases, there is a rapid growth in research and development in the medical sector. Even today, the data of the individual are stored manually in the medical sector, and it becomes tiresome and cumbersome to trace the user data. In turn, the doctor does not have the data instantly, and in the bargain, the patient is asked to undergo all the basic tests again, which is time consuming and leaves a hole in the pocket of the individual. Along with this, there is no track of drug manufacturing and distribution in the market, which leads to the rise in fake medicines. One can examine that there is no correlation among doctors, patients, drug manufacturers, basic testing labs, and so on. Since no one has control over the existing data, the security aspect becomes very significant. In spite of the implementation of latest devices such as laptops, cell phones, and so on in healthcare facility centers nowadays, one cannot assemble, investigate, protect, or interchange the facts impeccably. Hence, the healthcare system in today's scenario needs an unconventional system that can be incorporated smoothly, transparently, and economically with the existing system [15, 16]

Listed below are a few of the challenges in the blockchain technology in healthcare industry. There are numerous technical and administrative challenges that constrain the use of blockchain technology in the healthcare industry:

1. *Ambiguity*: As the development in the sector of blockchain concept is novel and ongoing, there are not many efficacious models based on which other models could be analyzed and processed. This becomes a major obstacle for rapid development.
2. *Storage Capability*: In the health sector, it is desirous of keeping the health records of all the individuals with parameters encompassing medical records, images, documents, lab reports, and so on. This will require the blockchain within the healthcare industry to have large space to store that amount of information.

3. *Data Ownership*: Ownership of the designed system in the healthcare sector is still unanswered. Some central body or the department has to be given authoritative rights so that some sense of ownership is developed.
4. *Capital*: As the structure and the network are massive and the data involved are also large, the cost and capital involved in designing and maintaining the structure are very unpredictable.
5. *Guideline Procedures*: There are no set of recommended guidelines that are readily available to be enforced on the blockchain in the healthcare industry. Although currently accountable regulatory body HIPAA (Health Insurance Portability and Accountability Act of 1996, United States) provides data privacy and security provisions, it is yet to be observed that how new guidelines concerning healthcare blockchain will adapt to existing privacy protocols (http://compare-trials.org).

8 Advantages of Blockchain

This helps in effective data tracking: rationalized and trouble-free information is shared between all the healthcare service providers. Their role is to provide data connectivity by offering reasonable cost-effective treatments and accurate analysis and ultimately healthier treatment for numerous illnesses. By exploiting and exploring the blockchain technology, all the healthcare service providers can work concurrently with the help of networks with enabled shared access. Real-time data tracking and authoritative security are the most important features of blockchain technology. Standardization, quickness, class, and security are other key benefits of blockchain technology.

1. *Authoritative monitoring*

As the network is gigantic and the data are continuously transmitted in real time, the security aspect becomes very important. This can be done effectively by guaranteeing a better healthcare management network with stable and authoritative monitoring. As an accurate response, the blockchain technology allows for documenting the transactions in a distributed record. It augments the accuracy and brings clarity and at the same time saves vital resources like time, costs, and efforts.

2. *Extracting the Best Benefits*

The technology is growing rapidly, and the network is carrying huge data in real time. A decision has to be made to collect and extract the finest benefits out of this healthcare data without thwarting the existing procedures.

9   Conclusion

Blockchain is a developing field and has a tremendous potential to affect a couple of enterprises, as well as it can change the manner in which businesses are done. This incorporates pharmaceutical organizations as well as clinics and healthcare institutions where the utilization of blockchain would prompt quicker and solve problems more efficiently. The reception of blockchain innovation in healthcare has begun, and we can hope to have business in blockchain arrangements in the market in not-so-distant future. A large portion of cases using blockchain in healthcare is intended to give patients a protected and integrated care.

From the above discussion, it is also very clear that the future of blockchain for the health and medicinal sector totally relies on the amalgamation and compliance of the new technology with the existing ecology. This will create a balanced technical structure. It has been observed that there are certain apprehensions concerning blockchain's amalgamation with current systems and its social acceptance, but still, this technology is favored in this sector.

References

1. Alhadhrami Z, Alghfeli S, Alghfeli M, Abedlla JA, Shuaib K (2017) Introducing block-chains for healthcare. In Proceedings of the 2017 International Conference on Electrical and Computing Technologies and Applications (ICECTA), Ras Al Khaimah, UAE, 21–23 Nov 2017; pp. 1–4. (CrossRef)
2. Angraal S, Krumholz HM, Schulz WL (2017) Blockchain technology: applications in health care. Circ Cardiovasc Qual Outcomes 10:e003800. (CrossRef) (PubMed)
3. Katuwal GJ, Pandey S, Hennessey M, Lamichhane B Applications of blockchain in healthcare: current landscape & challenges
4. CASP UK (2017) CASP checklists-CASP-critical appraisal skills programme. Qual Res Checkl 31:449
5. Eckblaw A, Azaria A, Hamalka J, Lippman A (2016) A case study for blockchain in healthcare: "MedRec" prototype for electronic health records and medical research data (White Paper)
6. Agrawal R, Chatterjee JM, Kumar A, Rathore PS (2020) Blockchain technology and the Internet of things: challenges and applications in bitcoin and security. Apple Academic Press. https://books.google.co.in/books?id=FCoMEAAAQBAJ
7. European Coordination Committee of the Radiological (2017) Block chain in healthcare; Technical report; European Coordination Committee of the Radiological: Brussels, Belgium
8. Evans M (2017) Why some of the worst cyberattacks in health care go unreported: Some breaches at hospitals involving ransomware don't have to be made public, a loophole some are trying to close. Wall Street J, 18 June 2017, p. B1
9. Health Affairs Health Policy Brief. Interoperability; 2014. https://doi.org/10.1377/hpb20140811.761828
10. Henry J, Pylypchuk Y, Searcy T, Patel V (2016) Adoption of electronic health record systems among U.S. Non-Federal Acute Care Hospitals. ONC Data Brief 2008–2015, 35
11. hospital-ehr-adoption-2008-2015.php. Accessed 26 Mar 2018
12. http://blog.appliedinformaticsinc.com/blockchain-transparency-integrity-and-clarity-to-the-healthcare-system
13. http://compare-trials.org

14. https://dashboard.healthit.gov/evaluations/data-briefs/non-federal-acute-care-
15. https://hackernoon.com/blockchain-in-healthcare-opportunities-challenges-and-applications
16. https://www.engineerbabu.com/blog/blockchain-in-healthcare-opportunities-challenges-and-applications
17. https://www.healthit.gov/sites/default/files/14-38-blockchain_medicaid_solution.8.8.15.pdf
18. https://www.ncbi.nlm.nih.gov/pmc/articles/PMC4105729/
19. https://www.usfhealthonline.com/resources/healthcare/does-blockchain-have-a-future-in-healthcare/
20. https://www2.deloitte.com/content/dam/Deloitte/us/Documents/public-sector/us-blockchain-opportunities-for-health-care.pdf
21. Kuo TT, Kim HE, Ohno-Machado L (2017) Blockchain distributed ledger technologies for biomedical and health care applications. J Am Med Inform Assoc 24:1211–1220. (CrossRef) (PubMed)
22. Leo Scanlon, deputy chief information security officer for the U.S. Department of Health and Human Services, before the U.S. Congress, House Energy and Commerce Committee. (2017, June 8). Washington, DC
23. Mackey TK, Nayyar G (2017) A review of existing and emerging digital technologies to combat the global trade in fake medicines. Expert Opin Drug Saf 16:587–602. (CrossRef) (PubMed)
24. Miliard M (2017) Blockchain's potential use cases for healthcare: hype or reality? Blockchain Technology for Healthcare: Facilitating the Transition to Patient-Driven Interoperability, William J. Gordon, Christian Catalin
25. Rosenbaum L (2015) Transitional Chaos or enduring harm? The EHR and the disruption of medicine. N Engl J Med 373:1585–1588. https://doi.org/10.1056/NEJMp1509961
26. Shackelford SJ, Myers S (2017) Block-by-block: leveraging the power of block chain technology to build trust and promote cyber peace. Yale J Law Technol 19:334–388
27. Shanafelt TD et al (2016) Relationship between clerical burden and characteristics of the electronic environment with physician burnout and professional satisfaction. Mayo Clin Proc 91(7):836–848
28. The U.S. Department of Defense Advanced Research Projects Agency (generally referred to as DARPA) developed the first packet switching network—a digital networking method of communications that groups transmitted data into blocks, called packets—and the first network to implement the TCP/IP protocol. Both technologies became the technical foundation of the Internet
29. U.S. Department of Health and Human Services (HHS) (2013) Summary of the HIPAA security rule. HHS, Washington, DC
30. Wüst K, Gervais A (2017) Do you need a Blockchain? IACR Cryptol ePrint Arch 2017:375

Blockchain-Based IoT Architecture

Shweta Sharma, Astha Parihar, and Kusumlata Gahlot

1 Introduction

In the Internet of Things (IoT) objects are associated with the internet with the assistance of a particular organization and sensors. These items associated with the internet communicate with one another or with bigger frameworks and trade data. Thus, they perform practically any undertaking you can consider. Although the set of experiences is old, the IoT has arrived at using the vital point improvement of innovation (Fig. 1).

Regions of use are growing step by step. Areas of smart homes, smart cities, in the Industry IoT, in IoT for wellbeing, not in instruction, not in horticulture, and many more are probably going to run over. Through the arrangements it proposes in savvy urban areas, it discovers answers for some issues, for example, transportation, natural contamination, and energy assets at the end-of-life. In brilliant institutions, normal work is finished through substances. Regardless of whether you're at home or not, the house is levelled out. It gives efficiency, profitability, and excellence in business life. Numerous arrangements in the wellbeing and clinical fields have been discovered thanks to IoT innovation. At the end of the day, zeroing in regarding this matter and leading examination will carry the future to a vital point. These days, IoT innovation is an answer to some money issues [1].

S. Sharma (✉) · A. Parihar · K. Gahlot
MDSU Ajmer, Ajmer, India

© Springer Nature Switzerland AG 2022
P. Raj et al. (eds.), *Blockchain, Artificial Intelligence, and the Internet of Things,*
EAI/Springer Innovations in Communication and Computing,
https://doi.org/10.1007/978-3-030-77637-4_10

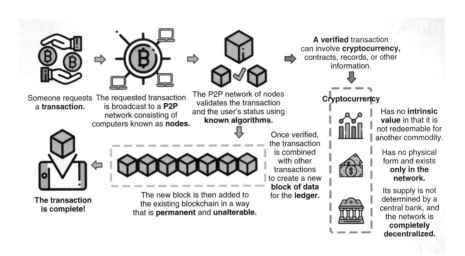

Someone requests a **transaction**.

The requested transaction is broadcast to a **P2P** network consisting of computers known as **nodes**.

The P2P network of nodes validates the transaction and the user's status using **known algorithms.**

A **verified** transaction can involve **cryptocurrency,** contracts, records, or other information.

Once verified, the transaction is combined with other transactions to create a new **block of data** for the **ledger.**

The transaction is complete!

The new block is then added to the existing blockchain in a way that is **permanent** and **unalterable.**

Cryptocurrency

Has no **intrinsic value** in that it is not redeemable for another commodity.

Has no physical form and exists **only in the network.**

Its supply is not determined by a central bank, and the network is **completely decentralized.**

Fig. 1 Internet of Things

2 Blockchain

It is the shared blockchain, that unchanging record inspires the way to recording exchanges and following resources in a business organization. A resource can be material (a house, vehicle, money, land) or immaterial (licensed innovation, licenses, copyrights, marking). Anything of significant value can be followed and exchanged on a blockchain network, decreasing danger and reducing expenses for all included. The information base is called the blockchain, which makes it conceivable to keep information records that increase constantly and make them a strong chain. It is a significant structure that is thought to achieve changes in the monetary area and numerous different areas. The innovation that frames the premise of the possibility that prompts significant victories, for example, bitcoin is blockchain innovation technology (Fig. 2).

The quicker it is obtained and the more precise it is, the better. Blockchain is ideal for conveying that data as it gives prompt, shared, and straightforward data hidden away on a changeless record that can be obtained exclusively by network individuals with permission. A blockchain organization can follow orders, instalments, records, creation, and significantly more. Furthermore, because individuals share a solitary perspective on reality, you can see all the subtleties of an exchange from start to finish, giving you more prominent certainty, just as new efficiencies and openings [2] (Fig. 3).

The data structure in blockchain can be named in a linked list. It creates the linked list as an unbreakable chain that does not allow various blocks to enter or exit. The chain adds to this condition. All that is to this requirement and all participants perform authentication. Blockchain advantages of the structure apart and the security of the private transaction are authorized participants to see the transactions

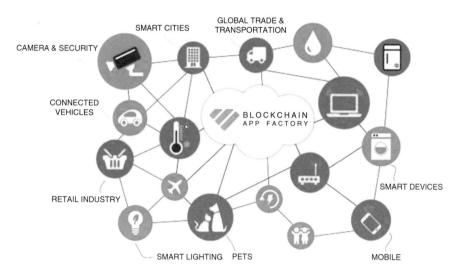

Fig. 2 Blockchain App Factory

in the chain. It defines the safe structure. It is possible to expand the database. Then, it is almost impossible to make changes to historical information. Thanks to this knowledge, end-to-end messaging has become secure. The best example is bitcoin (Fig. 4).

The idea of crypto currency and blockchain innovation, which became well known through bitcoin, has numerous different uses. This innovation is specifically compelling for those looking for a safe data set. Some of the uses are shown below: in the monetary business, particularly those looking for creative arrangements in banking are utilized in territories, for example, installment preparing, cash moves, character the board, etc. It is additionally utilized in certain regions to give inventive arrangements by using innovation in the public arena. It tends to be utilized democratically, making shrewd agreements, giving energy dispersion, providing energy distribution and assessment frameworks, etc. It can likewise be utilized in bigger tasks by consolidating with the web of objects [3] (Fig. 5).

2.1 Blockchain and Internet of Things

Innovation the blockchain is supposed to give incredible preferences through disposing of the safety issue in IoT innovation. The IoT will have the option to efficiently secure their gadgets against outside perils. These perils can be dispensed with through protected informing. For a third position without the requirement, the gadgets can collaborate securely through blockchain innovation. This framework assumes a significant part in following numerous items associated with the web. IoT blockchain is a significant idea for protected tasks. Blockchain companies in the IoT

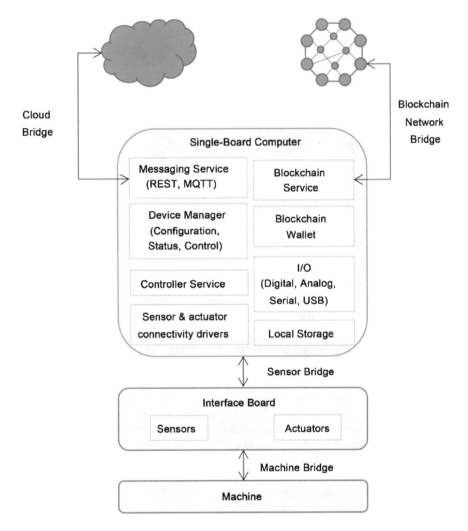

Fig. 3 Blockchain platform

appear in complete zones and provide limitless advantages. With the improvement due to innovation, advances are consolidated for more secure and more inventive arrangements. The blockchain of utilization innovation is in the IoT innovation. There are numerous organizations that do so. These portions of an organization are recorded:

1. California-based on a Helium San Francisco machine complex organization that utilizes blockchain in the IoT. It interfaces objects through the blockchain-based web. They have set up an effective blockchain framework and proceed to work.
2. Chronicled, located in San Francisco, California, this is another organization that utilizes blockchain innovation on the IoT. By applying blockchains to IoT

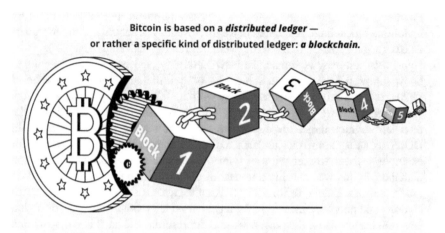

Fig. 4 Blockchain technology of the usage areas

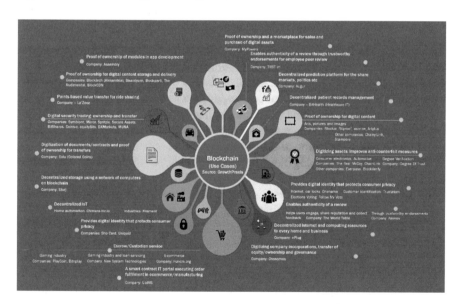

Fig. 5 Cases of the use of Blockchain technology

gadgets, the web of articles carries proficiency to the stock business with the assistance of extraordinary vehicle holders and sensors.

3. ArcTouch: the organization, situated in San Francisco, California, has built up a few blockchain applications that interface with IoT devices.

4. Filament: this organization, which produces equipment and programming by upholding blockchain, is situated in Reno, Nevada. The security of the IoT guarantees gadgets utilized in numerous ventures (energy, development, etc.).
5. Hypr: this company is situated in New York City, and utilizes blockchains to boost safety in the home and vehicle locks by utilizing the IoT knowledge of ATMs. It is a significant security worry with the IoT that has prevented its huge scope sending. IoT gadgets regularly endure security weaknesses that make them an obvious target for distributed denial of service (DDoS) assaults. In DDoS assaults, numerous undermined PC frameworks barrage a target, for example, a focal worker with an immense volume of concurrent information demands, in this way causing a refusal of administration for clients focused on the framework. Lately, various DDoS assaults have caused disturbance for associations and people. Unstable IoT gadgets provide an obvious target for digital lawbreakers to misuse the powerless security insurance to hack into dispatching DDoS attacks. Another issue of current IoT networks is adaptability. As the quantity of gadgets associated with an IoT network develops, currently incorporated frameworks to confirm, approve, and interface with various nodes in an organization will transform into a bottleneck. This would require an immense interest in workers who can deal with the huge quantity of data trade, and the whole organization can go down if the worker becomes inaccessible [4].

According to Gartner's Forecast, endpoints require the Web of Things to develop at a an annual development pace from 2016 through 2021 of 32%, arriving at an introduced base of units of 25.1 billion. Although the IoT gadgets are expected to become a particularly essential piece of our day-to-day lives in the coming years, it is a basic requirement that organizations put resources into tending to the above security and versatility challenges. For more detailed data on blockchain innovation, please refer to Deloitte's past distribution of the revolution of blockchain [5].

The number of developing blockchain conventions, associations, and IoT gadget suppliers currently demonstrates that there is a solid match for the blockchain area of the IoT. Some existing players and utilization of their cases are described briefly below [6]. (It would be ideal if you note that the associations and organizations referenced in this article ought not to be reflected as supported by Deloitte.) The chain of things (CoT) is a consortium of technologists and blockchain-driving organizations. It researches the most ideal use of situations where a combination of the IoT and blockchain can suggest critical advantages to mechanical, natural, and philanthropic presentations. Up until now, The CoT has assembled Maru, a coordinated IoT and blockchain equipment solution for addressing problems through personality, interoperability, and safety. Three use cases named chain of solar, chain of security, and chain of shipping were created. Particl is a convention for quick exchange expenditure and information respectability, with a tangle record that kills the requirement for costly mining (approval of exchanges) [7].

This innovation gets through the physical/advanced separation to find some kind of harmony among the paper documentation and the interest in favorable circumstances that the blockchain innovation needs to offer. Its modem joins sensors of the

IoT through blockchain innovation, giving honest information to exchanges including actual items. The natural situations of the modem sensors record, for example, a temperature that merchandise is dependent upon while in transit. At the point when the products arrive at the following travel point or end client, the sensor information is checked against preordained conditions in a smart agreement on the blockchain. The agreement approves that the surroundings meet all the prerequisites set out by the transmitter, its customers, or a controller, and generates different activities, for example, notices to sender and beneficiary, installment, or the arrival of goods. There are concerns regarding the reception of the IoT and blockchain advancements. As clarified already, a principal issue of existing IoT frameworks is security in their engineering, through a concentrated customer worker model overseen using a focal power which types it helpless to a solitary purpose of disappointment. This issue tends to blockchain through a decentralizing dynamic to an agreement based on the mutual organization of gadgets [8].

Nonetheless, while planning the design for IoT gadgets relating to a blockchain record, there are three principal difficulties to consider: scalability. One of the significant challenges is the scale of IoT –how to deal with the information gathered through enormous methods of a huge association of possibly and sensors lower exchange handling velocities or latencies. A reasonable description of the information model can already prevent challenges and save time while the arrangement is being established. The second difficulty is organizational safety and exchange classification. The security in the exchange history shared for a record association of IoT devices cannot be allowed on effectively open blockchains. On the grounds that design exchange of the investigation to make deductions can be applied almost the qualities of gadgets or clients behind open secrets. IoT and blockchain are both evolving advances with incredible potential, yet at the same time they lack wide selection because of specialization and security concerns. A few organizations in the current market areas use cases dealing with combining the two advances; together, they offer an approach to limiting security and going with corporate opportunities.

3   IoT-Based Security Architecture on Blockchain

The IoT is dependent on blockchain security design, as demonstrated. Our plan, essentially, IETF 6TiSCH, centers around the IoT. However, it can additionally apply to the next IoT security engineering comprising three sections: the border router, the blockchain network, and the IoT system (Fig. 6).

The IoT system framework comprises many hubs. Every hub is asset compelled (for example, restricted memory and force). Also, they do not have a place in a blockchain system. The hubs in the organization run a correspondence convention stack. The lower part utilizes the IEEE802.15.4 standard of convention stack, which is the ordinary short reach actual layer standard. The highest point of the convention stack embraces the standard constrained application protocol (CoAP), an

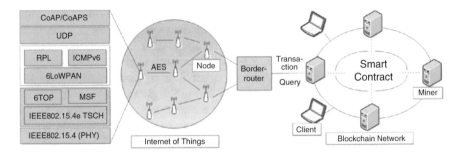

Fig. 6 Blockchain-based architecture of the IoT system

application that is a tiny convention like HTTP. In the IoT framework, hubs embrace ways of remote correspondence. As remote correspondence is not difficult to be spied upon, the correspondence ought encode messages through the advanced encryption standard (AES) algorithm [9].

3.1 Blockchain Network

A blockchain network mainly consists of many miners and customers. The border router periodically collects data uploaded by nodes into the IoT system and sends them to the blockchain network. Miners adopt a consensus algorithm (e.g., POW, POS). Leaders package some transactions into data blocks. Other miners can accept the data block after authenticating successfully. It is almost impossible to modify the data. The client can view the data at any time. Considering that most IoT systems can generate large amounts of sensor data, it is better to choose an advanced blockchain such as a private blockchain. Smart contracts are an important part of the blockchain network, allow automated transactions, and do not require third parties. Here, smart contracts can be used to access management. Only permitted customers can perform related operations, which can improve system security.

3.2 Border Router

The boundary switch assumes a significant part in the framework. To start with, it can deal with the asset-compelled hubs in the IoT framework. Only the hubs that are validated effectively can join the organization. Second, it assumes a transitional part. It interprets the sensor messages encoded through CoAP convention into the JavaScript object notation design satisfactorily using the blockchain hubs. Third, it can straightforwardly correspond with the miners, for example, send exchanges or questions. It ought to be furnished with a superior processor and huge memory, which is unique about the hubs in the IoT framework. The current Raspberry Pi can be used as a line switch. It can run a blockchain framework, for example, Ethereum. Security Analysis In this part, we provide a top to bottom security examination for the proposed design. Our security engineering gives complete consideration to the upside of blockchain innovation. It can give lots of private and security administration for the IoT framework. In contrast with the cloud-based focal framework, the blockchain network is a decentralization framework that has a preferred position in protecting against certain assaults (for example, denial-of-service [DOS] or DDoS assaults). Furthermore, the blockchain network stresses the issue of a solitary purpose of disappointment, which may occur in the cloud-based focal framework. The focal framework is generally controlled by an administrator. On the off-chance that the assailants take the records of the administrator, they can self-assertively alter the framework information. As we know, the information or interpretation in the blockchain network is altered opposition. Ensuring privacy is a significant security administration in the IoT framework. In the cloud-based focal framework, the client's information is hidden subjectively and is easily abused by programmers. The blockchain organization can give the self-governance administration using an open key cryptography component. Furthermore, the correspondence in the IoT framework embraces the AES encryption calculation, which is truly versatile according to the asset-compelled IoT gadget. The entrance control is additionally significant in the IoT framework; the smart agreement of the blockchain organization can provide this safety administration [10, 11].

4 Internet of Things for the Blockchain Architecture (Fig. 7)

Probably the greatest test in joining blockchain into the IoT is adaptability, owing to the enormous number of gadgets and asset limitations. The ideal blockchain design should scale to numerous IoT gadgets (they become the friends of the blockchain organization), and it should deal with a high throughput of exchanges. A combination of the IoT, the stage planned in this work, misuses both POW blockchains and Byzantine fault-tolerant (BFT) conventions to accomplish adaptability. To begin with, POW blockchains accomplish disseminated agreements among numerous IoT gadgets, their companions on the blockchain. At that point, Hybrid-IoT uses a BFT convention between connector structures to accomplish interoperability among the

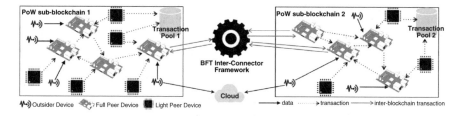

Fig. 7 Blockchain framework

sub-blockchains. To quantify and qualify the presentation of our methodology, we previously characterized a group of blockchain-IoT reconciliation measurements and tried our plan with a recreation system. We contemplated the affectability of blockchain boundaries, including blockchain block sizes and block age stretches, gadget areas, and several friends. The information accumulated from the huge number of recreations with countless mimicked gadgets was done on an IBM POWER8 supercomputer and encouraged us to outline this issue from the point of view of multi-target improvement. That is how to upgrade the grouping and determination of the sub-blockchains subject to versatility, security, companion's jobs, and other joining measurements. Assessment of these outcomes prompted the meaning of a group of "sweet spot" rules and a group of investigations that gather geologically appropriate IoT gadgets into sub-blockchains, as per inclinations made by the client (e.g., one could give more significance to security or versatility, and therefore apply various loads to the pertinent expense work) [11].

5 Centralized IoT Architecture

Dealing with a group of hubs to cooperate to configure a framework needs to have a specific engineering plan. Among the famous plans is the centralized design, which is fabricated utilizing a unified worker to control and deal with a group of hubs. These hubs shift from a high-level PC framework, PC, cell phone, and so on, which are equipped for performing different sorts of tasks. The incorporated worker functions as the supervisor who manages all solicitations coming from different hubs and oversees task planning and distribution among the hubs in the organization. This is straightforward type of concentrated design, where all hubs in the organization are associated through a focal worker [12] (Fig. 8).

6 Centralized Architecture

The IoT framework is among the regular instances of a concentrated framework, which is additionally called customer worker engineering. In this methodology, all IoT gadgets and items are associated, overseen, and validated through a centralized worker, who is normally a cloud worker. As stated, the incorporated IoT design comprises three key layers: the discernment, the application layer, and the organization. Even though there are a few concentrated engineering plans for the IoT framework comprising four, five, and six layers proposed by various analysts, this three-layer design shows the usefulness of the IoT framework easily and without any problems. It is a straightforward type of a unified design, where all hubs in the organization are associated through a focal worker the[13] (Fig. 9).

The principal layer of IoT engineering is the insight layer (also named the detecting layer). This layer includes different sorts of radiofrequency identifications,

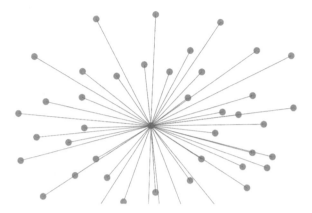

Fig. 8 Centralized IoT architecture

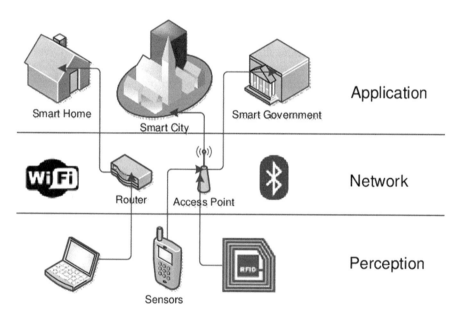

Fig. 9 Centralized architecture

sensors, wireless sensor networks, and actuators. The primary duty of this layer is to detect and see the general climate, gather the pertinent information that can be prepared, and extricate significant data to comprehend and devise our surrounding real world. In light of the gathered information, IoT gadgets can settle on setting mindful and self-sufficient choices utilizing actuators. The organization layer is used to associate and impart all IoT things and gadgets over the internet, where the concentrated worker is found. This layer consolidates passages that address the correspondence focuses among the discernment and organization layer. Different correspondence innovations and conventions were utilized in this layer, for example, ZigBee, 3G/4G, wi-fi, Bluetooth, and broadband to move the information between the discernment and the application layers. The application layer involves assorted IoT applications that use enormous quantities of information gathered and prepared at the insight and organization layer, individually, and produce digitized administrations in different areas, for example, medical care, keen stopping, smart home, brilliant city, wearables, smart lattice, agribusiness, and numerous others. The current model of the IoT framework provides a few points of interest to associate and impart a wide assortment of gadgets that are overseen by the concentrated worker. Consequently, the entire charge of the IoT network is overseen through a focal worker, which is less difficult to oversee and maintain [14]. Also, it saves the expense of actualizing a few complete workstations of equipment and programming in the organization, in which the greater part of the preparing activities are just taken care of by the concentrated worker. Thus, most hubs in the organization can resemble a terminal to interface with the focal worker. Plus, the concentrated IoT design conveys better actual security as most IoT information is kept in a solitary area, which is easier to shield from actual damage. On the other hand, the incorporated IoT engineering presents various difficulties. For instance, it faces versatility issues as it cannot deal with the steady expansion of IoT gadgets. Moreover, it presents various security and protection challenges, giving an outline of issues identified with the IoT centralized design [15] (Table 1).

7   IoT Architecture Represented by Four Building Blocks

The idea of the IoT is energizing and intriguing, though, one testing part of the IoT is a safe biological system as well as blocks of all structures in IoT design. The diverse block structure the understanding of IoT, zones of recognizing the weakness in each block, and investigating expected advances to counter is one of the fundamental shortcomings in IoT management of security issues (Fig. 10).

8   IoT Architecture Can Be Represented By Four Building Blocks

Things these are characterized as remarkably recognizable hubs, principally sensors that convey without human collaboration utilizing diverse availability strategies.

Gateways these are intermediaries between things and the cloud to give the required availability, security, and sensitivity.

Table 1 IoT centralized architecture

Challenges	Description
Single point of failure	As the unified worker carries out all preparation activities and oversees correspondence between different gadgets, there is a solitary disappointment where if the worker fails, the entire organization of gadgets will be inaccessible
Security	There is security between the vital difficulties in the IoT unified model, as entirely information handling tasks and information stockpiling are done in one area and through a focal worker, which makes it defenseless to various sorts of dangers, explicitly denial of service
Confidentiality	Different sorts of ongoing information including sensitive data are gathered from IoT gadgets, for example, propensities, passwords, individual and monetary data. This gathered information is reserved under the area fully controlled by the unified outside worker, who can abuse the information security. Furthermore, hiding it away in one area can make it easier to penetrate
Inflexibility	The concentrated worker controls interchanges and preparation activities between all hubs connected to the IoT organization, which leaves an enormous remaining task. To deal with this remaining task, the incorporated worker designs the heap to sidestep top burden issues. Nevertheless, this cutoff points to client adaptability while taking care of their undertakings because of the tight planning and postponement of being connected to this method
Cost	Performing all the central server preparation and correspondence tasks between all hubs in the organization requires high equipment and programming abilities to deal with this outstanding burden. Furthermore, gigantic holding stockpiles are required that can store information coming from different IoT gadgets. All these high capacities of equipment and programming are accompanied a significant expense
Scalability	Among the highest difficulties related to the centralized model is versatility. Overseeing and controlling all hubs in the organization being done by a focal worker can scale down well to smaller organizations. Utilizing the thought of a centralized framework with enormous venture associations that include a few branches in various territories would be nonsensical. The number of IoT gadgets is expanding continually, which implies that the incorporated model cannot scale and capacity productively
Access and diversity	Among the significant parts of a productive framework is the capacity to furnish admittance to every one of their clients with different necessities. Nevertheless, the centralized framework requires its clients to get to the data consistently utilizing indistinguishable cycles. Moreover, most centralized frameworks use a specific working framework for the entire organization, which confines variety to within the organization. For the IoT framework that contains heterogeneous and different gadgets, this will deliver a significant issue that requires being taken care of

Network organization this includes switches, aggregators, doors, repeaters, and different gadgets that control and secure the stream of information.

Cloud organization cloud foundation comprises huge pools of virtual workers and capacities that are organized along with registering and scientific abilities [16].

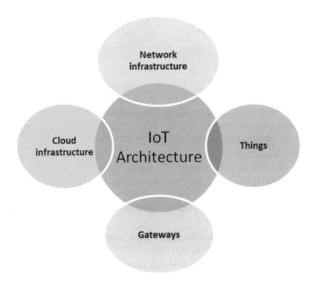

Fig. 10 Building blocks of IoT

8.1 Decentralizing IoT Networks

It is a decentralized way of dealing through IoT systems would administration address a number of the problems above. Shared implementation of a normalized handling of the correspondence model the billions of gadgets between exchanges will essentially diminish the expenses introducing related to maintaining and enormous concentrated will circulate calculation and server farms and needs across the capacity billions of devices that structure IoT organization. This will prevent disappointment in any single hub in an organization from bringing the whole organization to a complete breakdown [17].

Conventional IoT elements play out the arrangements without a unified device; any decentralized methodology should uphold three primary abilities:

* Distributed file-sharing
* Autonomous device coordination
* Peer-to-peer messaging

8.2 Approach of a Blockchain Technology

Blockchain, the "disseminated record" innovation, has arisen as an object of extreme interest in the tech business and in the past. Blockchain innovation offers a method of recording exchanges or any computerized association in a manner that is intended to be secure, straightforward, exceptionally impervious to blackouts, auditable, and productive; accordingly, it conveys the chance of disturbing ventures and empowering new plans of action. The innovation is youthful and changing quickly; inescapable commercialization is as yet a couple of years away. Nevertheless, to stay away from troublesome astonishments or botched freedoms, tacticians, organizers, and leaders across enterprises and business capacities should take notice now and start to explore utilization of the innovation [18].

8.3   *The Blockchain and the IoT*

In the IoT the missing connection to concerns about unwavering quality and settled protection is the blockchain innovation. It could be the blockchain innovation of a silver projectile required through the IoT business. Its utilization tends to be the association of billions of devices, empowering the handling of interactions and coordination between gadgets; for the critical investment considered funds for IoT industry makers. This single-purpose decentralized methodology would remove the disappointment, making a stronger biological system for devices to run on. The utilization of blockchains through cryptographic calculations would make the information purchaser more private [19].

The self-sufficient, decentralized, and trustless blockchain has the abilities to turn into an ideal segment of a basic element of IoT arrangements. It is nothing that the unexpected advances of the IoT have immediately obtained by one of the early adopters of blockchain innovation.

An IoT organization, the blockchain of historical records, and the unchanging backdrop of brilliant devices. The empowers of this element working of self-ruling smart devices the requirement without for concentrated power. Hence, the blockchain paves the way for an progression of conditions with the IoT that were strikingly troublesome, or even difficult to actualize without the IoT.

For example, by using blockchain, the IoT can empower secure arrangements, trustless informing between gadgets in an IoT organization. In this model, it will treat the blockchain message skills between devices like monetary connections in a bitcoin network. To authorize message trading, devices will use smart agreements, which at that point model the two gatherings between the arrangement.

9   What Are the Challenges? (Fig. 11)

The entirety of its advantages notwithstanding, the blockchain model is not without its flaws and inadequacies:

1. Versatility problems relating to the prompt centralization on the blockchain, a projection that is an eventual fate shadow over digital money.
2. The time needed and handling power to achieve encryption for all the associated articles through a biological system based on blockchain. IoT environments are extremely different. In contrast to non-exclusive processing organizations, IoT networks include different device-registering abilities, and not all of them will be fit for running related encryption calculations at the ideal speed.
3. Absence of abilities: barely any individuals see how blockchain innovation truly functions and when IoT is added to the blend that number will definitely be reduced.
4. Legitimate and consistency matters: it is a new region in all perspectives with no lawful or consistent code to follow, which is a difficult problem for manufactur-

Fig. 11 The IoT and blockchain challenges

ers and specialist co-ops. This will test and frighten away administrations from applying numerous blockchain improvements.

The ideal stage

The IoT creates answers for the required phenomenal joint determination, coordination, and availability for each piece in the environment, and entirely through the biological system overall. All devices should be coordinated and cooperate with any remaining gadgets, and all devices should interface and impart consistently through related structures and foundations [20].

9.1 The Optimal Platform for the IoT

The optimal platform for the IoT can:

1. Obtain and oversee information to make a guidelines-based, adaptable, and protected stage.
2. Secure and coordinate information to decrease total unpredictability although ensuring speculation.
3. Examine information and act by extricating business esteem from information, and afterward following up on it.
4. Security should be worked in as the establishment of the IoT situation, with thorough legitimacy checks, confirmation, information confirmation, and all the information is required to be scrambled.
5. Programming improvement and the application level of suggestions should be better at comprising code that is tough, reliable, secure, and through better code advancement norms, organizing, testing, and danger examination.

As frameworks connect, it is a fundamental requirement to have an agreed interoperability standard that is protected and legitimate. Without building a strong base top we will face more threats with each gadget that is added to it. What we need is a safe IoT with security ensured. Blockchain innovation is an attractive option if we can beat its pitfalls.

10 Conclusion

Protection and security problems have consistently hampered the advancement of the IoT. The blockchain is viewed as a capable innovation for progressing framework security. Security engineering for the IoT is dependent on blockchain. The asset-compelled sensor hub in the IoT sends the gathered information to the boundary switch, and they do not have a place in blockchain organizations. The boundary switch can easily correspond with the miners, for example, sending exchanges or questions. Smart agreements are utilized to assemble the entrance switch instruments. Security examination shows that the proposed security engineering can shield against a number of IoT assaults [21]. Its determination fabricates a genuine blockchain framework to confirm the possibility and adequacy of the proposed security engineering. Blockchain and IoT advancements can assist us with building a trusted, self-coordinated, open, and environmental IP insurance framework, which can include all the various gatherings in the IP insurance and exchange methods, and even they may not confide in one another. To the most awesome aspect of our insight, this is the main work that applying blockchain and IoT innovations on customary IP security and exchange environment. Our proposed strategy utilizes IoT innovation and blockchain record rather than manual chronicling and checking, which can decrease human mediation as much as could be expected. We have recognized the central issues where blockchain innovation can help to improve IoT applications. An assessment has likewise been provided to demonstrate the possibility of utilizing blockchain hubs on IoT gadgets.

References

1. Ericsson mobility report: on the pulse of the networked society, Ericsson, Tech. Rep. November 2017. [Online]. Available: http://www.ericsson.com/mobility-report
2. Atlam HF, Alenezi A, Alassafi MO, Wills GB (2018) Blockchain with the internet of things: benefits, challenges, and future directions. Int J Intell Syst Appl 10:40–48
3. Atlam HF, Walters RJ, Wills GB. Intelligence of things: opportunities & challenges. In Proceedings of the 2018 3rd cloudification of the Internet of Things (IoT), Paris, France, 2–4 July 2018, pp. 1–6
4. Atlam HF, Wills GB (2019) Intersections between IoT and distributed ledger. In: Advances in organometallic chemistry, vol 60. Elsevier BV, Amsterdam, pp 73–113
5. Atlam HF, Wills GB (2019) IoT security, privacy, safety, and ethics. In: Intelligent sensing, instrumentation and measurements. Springer Science and Business Media LLC, Berlin, pp 123–149
6. Atlam HF, Wills GB (2019) Technical aspects of blockchain and IoT. In: Advances in organometallic chemistry, vol 60. Elsevier BV, Amsterdam, pp 1–39
7. Conoscenti M, Vetro A, De Martin JC. Peer to peer for privacy and decentralization in the internet of things. In Proceedings of the 2017 IEEE/ACM 39th International Conference on Software Engineering Companion (ICSE-C), Buenos Aires, Argentina, 20–28 May 2017, pp. 288–290
8. Draft draft-IETF-core-comi-00, January 2017, work in progress. [Online]. Available: https://datatracker.ietf.org/doc/html/draft-ietf-core-comi-00
9. Fernandez-Carames TM, Fraga-Lamas P (2018) A review on the use of blockchain for the internet of things. IEEE Access 6:32979–33001
10. Wood G (2013) Ethereum: a secure decentralised generalised transaction ledger. [Online]. Available: http://gavwood.com/paper.pdf
11. Hugoson MÅ (2008) Centralized versus decentralized information systems: a historical flashback. IFIP Adv Inf Commun Technol 303:106–115
12. Yin S, Lu Y, Li Y. Design and implementation of IoT centralized management model with linkage policy. In Proceedings of the Third International Conference on Cyberspace Technology (CCT 2015), Beijing, China, 17–18 October 2015, pp. 5–9

13. Driscoll K, Hall B, Sivencrona H, Zumsteg P (2003) Byzantine fault tolerance, from theory to reality. In International Conference on Computer Safety, Reliability and Security (SAFECOMP03), pp. 235–248. [Online]. Available: http://citeseer.ist.psu.edu/696238.html
14. He K, Chen J, Du R, Wu Q, Xue G, Zhang X (2016) Deypos: deduplicatable dynamic proof of storage for multi-user environments. IEEE Trans Comput 65(12):3631–3645
15. Karafiloski E, Mishev A. Blockchain solutions for big data challenges: a literature review. In Proceedings of the IEEE EUROCON 2017 – 17th International Conference on Smart Technologies, Ohrid, Macedonia, 6–8 July 2017, pp. 763–768
16. Der Stok PV, Bierman A, Veillette M, Pelov A. "CoAP management interface", Internet Engineering Task Force, Internet
17. Reyna A, Martín C, Chen J, Soler E, Díaz M (2018) On blockchain and its integration with IoT. Challenges and opportunities. Future Gener Comput Syst 88:173–190
18. Agrawal R, Chatterjee JM, Kumar A, Rathore PS (2020) Blockchain technology and the internet of things: challenges and applications in bitcoin and security, Apple Academic Press. https://books.google.co.in/books?id=FCoMEAAAQBAJ
19. Dziembowski S, Faust S, Kolmogorov V, Pietrzak K. "Proofs of space". In Advances in cryptology – CRYPTO 2015 – 35th Annual Cryptology Conference, Santa Barbara, CA, USA, 16–20 August 2015, Proceedings, Part II, 2015, pp. 585–605. [Online]. Available: https://doi.org/10.1007/978-3-662-48000-7
20. Sun X, Ansari N (2016) Edgeiot: mobile edge computing for the internet of things. IEEE Commun Mag 54(12):22–29
21. Shelby Z, Hartke K, Bormann C. "The Constrained Application Protocol (CoAP)," RFC 7252 (Proposed Standard), Internet Engineering Task Force, June 2014, updated by RFC 7959. [Online]. Available: http://www.ietf.org/rfc/rfc7252.txt\

Index

© Springer Nature Switzerland AG 2022
P. Raj et al. (eds.), *Blockchain, Artificial Intelligence, and the Internet of Things*,
EAI/Springer Innovations in Communication and Computing,
https://doi.org/10.1007/978-3-030-77637-4

Printed in the United States
by Baker & Taylor Publisher Services